# Capturing Upside Risk

Finding and Managing Opportunities in Projects

# Capturing Upside Risk
Finding and Managing Opportunities in Projects

David Hillson

CRC Press
Taylor & Francis Group
Boca Raton London New York

CRC Press is an imprint of the
Taylor & Francis Group, an **informa** business

AN AUERBACH BOOK

Microsoft® and Microsoft® Excel® are registered trademarks of the Microsoft Corporation.

*PMBOK® Guide*, PMI®, and PMI-RMP® are registered marks of the Project Management Institute, Inc., which is registered in the United States and other nations.

CRC Press
Taylor & Francis Group
6000 Broken Sound Parkway NW, Suite 300
Boca Raton, FL 33487-2742

© 2019 by Taylor & Francis Group, LLC
CRC Press is an imprint of Taylor & Francis Group, an Informa business

No claim to original U.S. Government works

Printed on acid-free paper

International Standard Book Number-13: 978-0-8153-8251-5 (Hardback)

This book contains information obtained from authentic and highly regarded sources. Reasonable efforts have been made to publish reliable data and information, but the author and publisher cannot assume responsibility for the validity of all materials or the consequences of their use. The authors and publishers have attempted to trace the copyright holders of all material reproduced in this publication and apologize to copyright holders if permission to publish in this form has not been obtained. If any copyright material has not been acknowledged please write and let us know so we may rectify in any future reprint.

Except as permitted under U.S. Copyright Law, no part of this book may be reprinted, reproduced, transmitted, or utilized in any form by any electronic, mechanical, or other means, now known or hereafter invented, including photocopying, microfilming, and recording, or in any information storage or retrieval system, without written permission from the publishers.

For permission to photocopy or use material electronically from this work, please access www.copyright.com (http://www.copyright.com/) or contact the Copyright Clearance Center, Inc. (CCC), 222 Rosewood Drive, Danvers, MA 01923, 978-750-8400. CCC is a not-for-profit organization that provides licenses and registration for a variety of users. For organizations that have been granted a photocopy license by the CCC, a separate system of payment has been arranged.

**Trademark Notice:** Product or corporate names may be trademarks or registered trademarks, and are used only for identification and explanation without intent to infringe.

**Visit the Taylor & Francis Web site at**
**http://www.taylorandfrancis.com**

**and the CRC Press Web site at**
**http://www.crcpress.com**

# Contents

| | |
|---|---|
| Contents | vii |
| List of Figures | xv |
| List of Tables | xix |
| Foreword | xxi |
| Preface | xxv |
| About the Author | xxix |
| **SECTION A—WHY OPPORTUNITIES MATTER** | **1** |
| **Chapter 1: What Are Opportunities?** | **3** |
|     First principles | 3 |
|         Uncertainty is everywhere | 4 |
|         Not all uncertainty matters | 4 |
|         Not all uncertainties that matter are bad | 5 |
|     Risk is "uncertainty that matters" | 6 |
|     The definition debate | 8 |
|     The great divide | 9 |
|     Reasons for reluctance | 16 |
|         Ignorance | 17 |
|         Language | 18 |
|         Culture | 18 |
|         Psychology | 21 |
|         Inertia | 22 |

| | |
|---|---:|
| **Winning the war** | 23 |
| Education and demonstration | 23 |
| Persistence | 24 |
| Memetics | 24 |
| Emotional intelligence | 25 |
| Support | 26 |
| **From theory to practice** | 27 |

## Chapter 2: What Are Opportunities in Projects?     29

| | |
|---|---:|
| **Why projects are risky** | 29 |
| Common characteristics | 30 |
| Deliberate design | 31 |
| External environment | 31 |
| **Synonym confusion** | 32 |
| **Links between opportunities and threats** | 33 |
| **Risks and risk** | 34 |
| **All projects include opportunities** | 36 |

## Chapter 3: How to Manage Project Opportunities     39

| | |
|---|---:|
| **Elements of a risk management process** | 40 |
| Nine structuring questions | 40 |
| Tailoring the risk process | 42 |
| **Use a common process for threats and opportunities** | 45 |
| Nine structuring questions for opportunities | 45 |
| Benefits of integration | 47 |
| **Introduction to Section B** | 48 |

## SECTION B—MANAGING PROJECT OPPORTUNITIES     51

## Chapter 4: Setting the Scene for Opportunity Management     53

| | |
|---|---:|
| **Purpose and principles of risk management planning** | 53 |
| Define objectives at risk and scope of risk process | 54 |
| Reflect risk appetite of key stakeholders in measurable risk thresholds | 55 |
| Tailor risk process to match the risk challenge of the project | 55 |
| Summary | 56 |
| **The typical risk management plan** | 56 |
| Purpose | 56 |
| Contents | 57 |
| **Understanding risk appetite and risk thresholds** | 63 |
| **Defining risk assessment criteria** | 65 |

| | |
|---|---:|
| Defining risk assessment framework | 70 |
| Summary and reflection questions | 73 |
|     Reflection questions | 73 |
| The next step ("Now we've defined the process, where are the opportunities?") | 74 |

## Chapter 5: Finding Opportunities     75

| | |
|---|---:|
| Purpose and principles of risk identification | 75 |
|     All risks are uncertain | 77 |
|     Each risk must be linked to at least one objective | 77 |
|     Consider different time perspectives | 78 |
|     Use more than one risk identification technique | 79 |
|     Include multiple perspectives | 79 |
|     Consider all potential sources of risk | 79 |
|     Repeat risk identification throughout the project | 80 |
|     Summary | 81 |
| Separating risk from non-risk using risk metalanguage | 81 |
| Typical techniques for identifying threats | 82 |
|     Past-focused techniques | 83 |
|     Present-focused techniques | 83 |
|     Future-focused techniques | 84 |
| Modifying threat techniques to find opportunities | 84 |
|     Modifying past-focused techniques | 84 |
|     Modifying present-focused techniques | 85 |
|     Modifying future-focused techniques | 89 |
| Using "two-dimensional techniques" | 90 |
|     SWOT Analysis | 91 |
|     Force-Field Analysis | 93 |
| Using risk metalanguage to identify opportunities | 95 |
| Separate or together? | 96 |
| A mindset for opportunity | 97 |
| Write it down | 99 |
| Summary and reflection questions | 100 |
|     Reflection questions | 101 |
| The next step ("Now we've identified the opportunities, which ones are most important?") | 101 |

## Chapter 6: Picking Winners     103

| | |
|---|---:|
| Purpose and principles of qualitative risk assessment | 103 |
|     Use consistent and objective assessment framework | 104 |

| | |
|---|---:|
| Reflect corporate risk appetite and risk thresholds | 105 |
| Seek input from different perspectives | 105 |
| Beware bias | 106 |
| Summary | 106 |
| **Defining prioritisation dimensions** | 106 |
| **Typical techniques for prioritising threats** | 107 |
| Two dimensions: P-I Matrix | 109 |
| More detailed prioritisation in the P-I Matrix | 111 |
| Three dimensions: Other approaches | 114 |
| **Typical techniques for categorising threats** | 116 |
| Using hierarchical project breakdown structures | 117 |
| Timing issues | 119 |
| **Modifying threat techniques to prioritise opportunities** | 121 |
| Adapting the P-I Matrix for opportunities | 122 |
| **Write it down** | 123 |
| **Summary and reflection questions** | 125 |
| Reflection questions | 126 |
| **The next step ("Now we've found the important opportunities, what can we do about them?")** | 126 |

## Chapter 7: Using Numbers to Model Opportunities — 129

| | |
|---|---:|
| **Purpose and principles of quantitative risk analysis** | 129 |
| QRA is not always needed | 130 |
| Where QRA is required, know why and how to use it | 131 |
| No models are "correct" | 132 |
| Include all types of uncertainty | 132 |
| Use best available data | 133 |
| Use the results | 133 |
| Summary | 134 |
| **Typical techniques for modelling threats** | 135 |
| Monte Carlo simulation | 135 |
| Generating input data | 137 |
| Understanding outputs | 146 |
| Using QRA results to evaluate overall project risk | 149 |
| **Modifying threat techniques to model opportunities** | 151 |
| **Write it down** | 156 |
| **Summary and reflection questions** | 157 |
| Reflection questions | 157 |
| **The next step ("Now we've quantified overall risk exposure, what can we do about it?")** | 158 |

## Chapter 8: Deciding What to Do — 159

### Purpose and principles of risk response planning — 159
- Strategy before tactics — 160
- Deal equally with both threats and opportunities — 161
- Respond to overall risk, not just individual project risks — 161
- The first idea is not always best — 161
- Responses must match the level of risk — 162
- Be creative but realistic — 162
- Ensure clear ownership — 163
- Summary — 163

### Typical techniques for responding to threats — 164

### Modifying threat techniques to respond to opportunities — 165
- Escalate — 166
- Exploit — 166
- Share — 167
- Enhance — 167
- Accept — 168

### Selecting preferred strategy — 169
- Cost-effectiveness — 171
- Risk-effectiveness — 172
- Timeliness — 173
- Secondary risks — 174

### Ensuring ownership — 175
### Turning strategy into actions — 176
### Write it down — 176
### Summary and reflection questions — 177
- Reflection questions — 178

### The next step ("Now we've planned responses to our opportunities, let's make sure we do them.") — 178

## Chapter 9: Taking Action — 179

### Purpose and principles of risk response implementation — 179
- Don't cut corners — 180
- Motivate action owners — 181
- Assume nothing — 181
- Summary — 181

### Just do it! — 182
### Write it down — 185
### Summary and reflection questions — 185
- Reflection questions — 186

The next step ("Now we've implemented our opportunity
responses, who should we tell and how?")  187

## Chapter 10: Telling Others   189

Purpose and principles of risk communication
and reporting   189
    Be honest   190
    Be specific   191
    Be timely   192
    Summary   192
Stakeholder risk information needs analysis   193
Risk communication design   194
Now communicate!   198
Delivery timeliness   198
Communicating about opportunities: together or separate?   199
    Risk register   199
    "Top risks" list   200
    Risk distributions   201
    Metrics and trend analysis   204
Summary and reflection questions   207
    Reflection questions   207
The next step ("Now we've told people about current risks,
how do we keep it up to date?")   207

## Chapter 11: Keeping Up to Date   209

Purpose and principles of risk reviews   210
    Decide when and where to review risk   212
    Review risk regularly   212
    Don't limit reviews to meetings   212
    Involve key stakeholders   213
    Summary   213
Typical risk review techniques   214
    Assess status of existing risks   214
    Review risk response effectiveness   217
    Identify new risks   217
    Review risk process effectiveness   218
Summary and reflection questions   218
    Reflection questions   219
The next step ("Now we've reached a key point in the
project, what have we learned?")   219

## Chapter 12: Identifying Risk-Related Lessons — 221

- Purpose and principles of identifying risk-related lessons — 221
  - Identify lessons from both bad and good experience — 223
  - Do it regularly, don't wait until project completion — 223
  - A lesson is not learned until it's implemented — 224
  - Record lessons in actionable form — 224
  - Summary — 224
- Typical techniques for identifying risk-related lessons — 225
- Write it down — 227
- Summary and reflection questions — 229
  - Reflection questions — 229
- Final reflection questions — 229

## SECTION C—AND FINALLY... — 233

## Chapter 13: Making It Work — 235

- Critical success factors for managing opportunities in projects — 236
  - Common risk language — 237
  - Simple scaleable risk process — 238
  - Appropriate supporting risk infrastructure — 239
  - Strong and mature risk culture — 241
  - Organisational learning — 245
  - CSF summary — 246
- Remaining work — 247
  - Standards and guidelines — 248
  - Process modifications — 248
  - Supporting risk software tools — 250
- The future for opportunity management — 252

## References and Further Reading — 255

- General references and further reading — 255
- Specific references for chapters — 257

## Index — 263

# List of Figures

| | | |
|---|---|---|
| Figure 1-1 | Q1. Organisational Definition of Risk | 11 |
| Figure 1-2 | Q2. Organisational Approach to Risk Management | 12 |
| Figure 1-3 | Q3. Personal Definition of Risk | 14 |
| Figure 1-4 | Q4. Support for Broader Risk Definition | 15 |
| Figure 1-5 | An Example of Maslow's "Hierarchy of Needs" | 21 |
| Figure 4-1 | Example Risk Breakdown Structure (RBS). (*Source:* Adapted from Hillson, D. A. (2009). *Managing risk in projects.* Farnham, UK: Routledge/Gower, with permission.) | 58 |
| Figure 4-2 | Risk Appetite Levels and Degree of Variation | 66 |
| Figure 4-3 | Defining Levels of Impact | 70 |
| Figure 4-4 | Probability-Impact Matrix for Threats | 71 |
| Figure 4-5 | Combined "Mirror" Probability-Impact Matrix for Threats and Opportunities. (*Source:* Reproduced from Hillson, D. A. (2009), with permission.) | 72 |
| Figure 5-1 | Cause-Risk-Effect Relationships. (*Source:* Adapted from Hillson, D. A. (2003). *Effective opportunity management for projects: Exploiting positive risk.* Boca Raton (FL), USA: Routledge/Taylor & Francis, with permission.) | 82 |
| Figure 5-2 | Example Benefit Tree | 88 |
| Figure 5-3 | SWOT Analysis Process. (*Source:* Reproduced from Hillson, D. A. & Simon, P. W. (2012). *Practical project risk management: The ATOM Methodology*, (second edition). Vienna (VA), USA: Berrett-Koehler, with permission.) | 92 |
| Figure 5-4 | Example Force-Field Analysis Diagram. (*Source:* Reproduced from Hillson, D. A. (2003), with permission.) | 94 |
| Figure 5-5 | Using Risk Metalanguage to Identify Opportunities | 95 |
| Figure 6-1 | Plotting Threats on the P-I Matrix | 110 |
| Figure 6-2 | Risk Scoring Using Two Linear Sequences 1/2/3/4/5 | 112 |
| Figure 6-3 | Risk Scoring Using Linear/Non-Linear Sequences. (*Source:* Adapted from Hillson, D. A. (2003), with permission.) | 113 |

| | | |
|---|---|---|
| Figure 6-4 | Example Bubble Chart (Urgency, Manageability, Impact). (*Source:* Adapted from Hillson, D. A. (2009), with permission.) | 115 |
| Figure 6-5 | Example Risk Prioritisation Chart (Probability, Impact, Proximity). (*Source:* Adapted from Hillson, D. A. (2009), with permission.) | 116 |
| Figure 6-6 | Example Risk Breakdown Matrix (RBS/WBS). (*Source:* Adapted from Hillson, D. A. (2003), with permission.) | 119 |
| Figure 6-7 | Overlapping Action Windows and Impact Windows. (*Source:* Reproduced from Hillson, D. A. & Simon, P. W. (2012), with permission.) | 121 |
| Figure 6-8 | Combined "Mirror" Probability-Impact Matrix (P-I Matrix) for Threats and Opportunities. (*Source:* Reproduced from Hillson, D. A. (2009), with permission.) | 123 |
| Figure 7-1 | Alternative Range Types for Modelling Variability. (*Source:* Adapted from Hillson, D. A. & Simon, P. W. (2012), with permission.) | 139 |
| Figure 7-2 | Example Probabilistic Branch—Acceptance Trials | 140 |
| Figure 7-3 | Example Probabilistic Branch—Planning Permission | 142 |
| Figure 7-4 | Example Conditional Branch—Outsourced or In-House | 143 |
| Figure 7-5 | Probabilistic Branch for Modelling a Threat into the Project Schedule | 144 |
| Figure 7-6 | Probabilistic Branch for Modelling a Threat into the Project Budget | 144 |
| Figure 7-7 | Example S-Curves for Project End Date and Cost | 147 |
| Figure 7-8 | Example Tornado Chart. (*Source:* Adapted from Hillson, D. A. (2003), with permission.) | 148 |
| Figure 7-9 | Example S-Curve for Total Project Cost. (*Source:* Reproduced from Hillson, D. A. (2009), with permission.) | 150 |
| Figure 7-10 | Example Probabilistic Branch with Positive Alternative Path | 154 |
| Figure 7-11 | Probabilistic Branch for Modelling an Opportunity into the Project Schedule | 155 |
| Figure 7-12 | Probabilistic Branch for Modelling an Opportunity into the Project Budget | 156 |
| Figure 8-1 | Selecting Risk Response Strategy by Position on P-I Matrix | 170 |
| Figure 8-2 | Prioritising Risk Response Strategies by Intensity. (*Source:* Adapted from Hillson, D. A. (2003), with permission.) | 171 |
| Figure 8-3 | Cost-Benefit Ratio Calculations for Threats and Opportunities | 172 |
| Figure 8-4 | Risk-Benefit Ratio Calculations for Threats and Opportunities | 173 |
| Figure 10-1 | Example Risk List Format (Prioritised). (*Source:* Reproduced from Hillson, D. A. (2003), with permission.) | 196 |
| Figure 10-2 | Summary Risk Report Example Contents List. (*Source:* Adapted from Hillson, D. A. (2003), with permission.) | 197 |
| Figure 10-3 | Detailed Risk Report Example Contents List. (*Source:* Adapted from Hillson, D. A. (2003), with permission.) | 197 |
| Figure 10-4 | Risk Distribution by Priority Category. (*Source:* Reproduced from Hillson, D. A. (2003), with permission.) | 202 |

Figure 10-5  Example Risk Metrics Data. (*Source:* Reproduced from Hillson, D. A. (2003), with permission.) 205
Figure 10-6  Risk Metrics and Trends (Data from Figure 10-5). (*Source:* Reproduced from Hillson, D. A. (2003), with permission.) 206
Figure 11-1  Risk Status Values. (*Source:* Adapted from Hillson, D. A. (2003), with permission.) 215
Figure 12-1  Example L2BL Register Format, with Sample Entries 228
Figure 13-1  Double Meaning of CSF 237
Figure 13-2  A-B-C Model for Risk. (*Source:* Adapted from Hillson, D. A. (2013). "The A-B-C of risk culture—How to be risk-mature." Proceedings of the PMI Global Congress North America 2013, New Orleans (LA), USA, 28 October 2013, with permission.) 242

# List of Tables

| | | |
|---|---|---|
| Table 1-1 | Definitions of Risk as "Uncertainty That Matters" | 7 |
| Table 1-2 | Risk Definition Survey Questions | 10 |
| Table 1-3 | National Characteristics Driven by Uncertainty Avoidance | 19 |
| Table 2-1 | Risks and Risk in Current Guidelines | 35 |
| Table 3-1 | Mapping the Nine Questions to International Risk Standards | 49 |
| Table 4-1 | Purpose and Principles of Risk Management Planning | 56 |
| Table 4-2 | Risk Data Held in the Risk Register | 60 |
| Table 4-3 | Risk-Related Roles and Responsibilities. (*Source:* Adapted from Hillson, D. A. (2009). *Managing risk in projects.* Farnham, UK: Routledge/Gower, with permission.) | 62 |
| Table 4-4 | Example Risk Assessment Criteria for Likelihood and Impact | 63 |
| Table 5-1 | Purpose and Principles of Risk Identification | 80 |
| Table 5-2 | Example Risk Descriptions Using Risk Metalanguage. (*Source:* Reproduced from Hillson, D. A. & Simon, P. W. (2012). *Practical project risk management: The ATOM Methodology*, (second edition). Vienna (VA), USA: Berrett-Koehler, with permission.) | 83 |
| Table 5-3 | Example Assumptions and Constraints Analysis Worksheet | 87 |
| Table 5-4 | Example Risks Derived from SWOT Analysis | 93 |
| Table 5-5 | Risk Data to Record after Risk Identification | 100 |
| Table 6-1 | Purpose and Principles of Qualitative Risk Assessment | 106 |
| Table 6-2 | Risk Prioritisation Factors | 108 |
| Table 6-3 | Example Risk Scoring Scheme with Non-Linear Impact Scale | 113 |
| Table 6-4 | Risk Data to Record after Qualitative Risk Assessment | 124 |
| Table 7-1 | Purpose and Principles of Quantitative Risk Analysis (QRA) | 134 |
| Table 8-1 | Purpose and Principles of Risk Response Planning | 163 |
| Table 8-2 | Generalising Threat Response Strategies to Deal with Opportunities. (*Source:* Adapted from Hillson, D. A. (2001). "Effective strategies for exploiting opportunities." Proceedings of the 32nd Annual Project Management Institute Seminars & Symposium (PMI 2001), Nashville (TN), USA, 5–7 November 2001, with permission.) | 165 |

| | | |
|---|---|---|
| Table 8-3 | Risk Data to Record after Risk Response Planning | 177 |
| Table 9-1 | Purpose and Principles of Risk Response Implementation | 181 |
| Table 9-2 | Risk Data to Record after Risk Response Implementation | 185 |
| Table 10-1 | Purpose and Principles of Risk Communication and Reporting | 192 |
| Table 10-2 | Common and Specific Project Stakeholders | 193 |
| Table 10-3 | Stakeholder Risk Information Needs Analysis, with Sample Entry. (*Source:* Adapted from Hillson, D. A. (2003). *Effective opportunity management for projects: Exploiting positive risk.* Boca Raton (FL), USA: Routledge/Taylor & Francis, with permission.) | 194 |
| Table 10-4 | Risk Outputs Definition, with Sample Entry. (*Source:* Adapted from Hillson, D. A. (2003), with permission.) | 198 |
| Table 11-1 | Purpose and Principles of Risk Reviews | 213 |
| Table 12-1 | Purpose and Principles of Identifying Risk-Related Lessons | 224 |
| Table 12-2 | Risk Process Steps—Purpose and Principles | 230 |
| Table 13-1 | Characteristics of Different Risk Attitudes | 244 |
| Table 13-2 | CSF Summary | 247 |

# Foreword

The value of any activity or project for an organisation is linked intrinsically to its objectives. Unless an activity or project enhances the organisation's purpose and objectives, then it has little value. David defines risk as 'uncertainty that matters'. Something matters if it has an impact on objectives, otherwise it doesn't matter and it represents no risk. Similarly, if there is no uncertainty then there's no risk, whether objectives are affected or not.

The relationship between activities and value is often complicated, and objectives provide a concise framework for linking the two. Several international standards adopt this approach to understanding how much risks matter and how risk management can help managers make better, value-adding decisions. ISO 31000:2018 *Risk management—Guidelines* makes explicit the relationship between risk management and value: 'The purpose of risk management is the creation and protection of value'. IEC 62198 *Managing risk in projects—Application guidelines* expands on this: 'Risk management contributes to the demonstrable achievement of objectives and improvement of performance and quality in projects and the assets, products and services they create'.

Despite this, much of the literature on risk management, and much of its practice, focuses solely on protecting value by reducing the likelihood and extent of negative outcomes on objectives. In the context of project risk management, this inevitably focuses attention on threats, the sources of risks that might have negative implications for project value, overlooking wider possibilities to achieve better outcomes.

The contribution of this book is to help readers move from a narrow focus on detrimental effects on objectives to a broader view of *all* effects on objectives. Uncertainty that matters, or risk, may be associated with either positive or negative effects on objectives, or both—the definition embraces all kinds of impacts. This supports the guiding principle in ISO 31000, that effective risk management *creates value*, as well as protecting it. Managing those sources of risk that may lead to positive impacts, or opportunities, forms an important part of getting the most from project risk management.

Although the title of this book includes the word 'opportunities', this is not its sole focus. Most of what David describes here applies to all project risk management, not just to those parts that deal with positive consequences for objectives and outcomes. He describes how to manage all kinds of project risks, both threats and opportunities, emphasising the broad interpretation of the definition of risk.

Nevertheless, there are aspects of the project risk management process as it is often undertaken that should be conducted slightly differently if its full value is to be realised. David explains the project risk management process, and he provides both the rationale for enhancing it and guidance on how to take full advantage of the wider perspective.

The book has three sections: three chapters dealing with why opportunities matter, a longer section on the process for managing opportunities in projects, and a final short section on making it work.

The second section of the book is structured to provide a step-by-step guide to the project risk management process, with a chapter devoted to each step. An initial discussion of the purpose and principles for each chapter sets out the scope and key points.

The chapter on finding opportunities is particularly valuable. It starts with well-known approaches for risk identification, based on the past (looking back), the present (looking around) and the future (looking forward). Most of these techniques are directed to identifying threats, sources of risk with negative impacts on objectives. But David goes on to discuss how past-focused, present-focused and future-focused techniques can be modified to find opportunities.

David points out that past-focused techniques, like those based on checklists and lessons learned, may be of little use until an organisation has built up historical resources and experience in identifying opportunities. However, present-focused techniques like influence diagrams and assumptions analysis are readily adapted for opportunities, and more specialised techniques like fault tree analysis can be adapted to construct benefit trees.

In the realm of future-focused techniques, brainstorming is one of the most common risk identification processes in practice. Risk managers and risk workshop facilitators will find the tips on enhancing brainstorming to better identify opportunities particularly useful. Thinking processes like 'three wishes', 'good luck', 'in your dreams' and 'what if not?' are all simple, straightforward and easy to apply, but they are not common tools for many facilitators. Pre-mortems, combining elements of futures thinking, scenario analysis and visualisation are also outlined.

Other 'two dimensional' tools that identify threats and opportunities at the same time are also discussed, including SWOT analysis and force field analysis.

A key section of the chapter on finding opportunities deals with developing a mindset for identifying opportunities. For many project teams and project risk practitioners, project risk management has been about threats and negative impacts, and it has been this way for many years. Breaking out of this mode of thinking requires active intent; David lists some of the defining characteristics of a mindset that is open to opportunities and offers advice on how to reinforce them.

This book is easy to read. The writing is relatively informal, but precise, and the style is clear, logical, and persuasive. The clarity of thought and expression explains why David is such a sought-after speaker. This book is a must-read for project risk managers, as well as for project professionals who are serious about addressing all the risks on their project, including the good ones.

Dr Dale F Cooper<br>
Director, Broadleaf Capital International<br>
Cammeray, Australia<br>
May 2019

# Preface

A month ago, my wife and I were on a sunrise game drive in South Africa. None of the animals seemed to be awake yet, and all was quiet. I sat on the edge of the vehicle, my camera round my neck and ready, more in hope than expectation. Suddenly and without warning, an adult cheetah emerged from the bush two metres from my elbow, looked straight at me, gave a ferocious snarl, and disappeared again. Click! I was so glad to be ready for the unexpected event, and to have a great photo to prove it.

In 1990 I was the risk specialist on a team taking part in a feasibility study, developing a gun fire control system. The British Army had booked the firing range and some tanks for a competitive shoot-off between systems from our company and one other. The trial date was fixed, the timeline was tight, and we had to cut some features in order to meet the deadline. But we maintained a wish-list of the functions we'd omitted, just in case we had the chance to put them in at some point. One week before the trial, the Army announced a two-week delay for operational reasons. We immediately remobilised our team, implemented several of the features at the top of our wish-list, did some additional system integration tests, and submitted an improved solution for the postponed trial. Meanwhile the competitor company had given their team a week off as a reward for working so hard to meet the original date. The trial took place, our system outperformed the other in the shoot-off, and we won the main development contract. Our "just-in-case" plan made all the difference.

What links these two disparate stories is *awareness* and *preparedness*. In both cases, we recognised the possibility that something unexpected might happen, and we were geared up for it. We had a nice surprise, and we were able to take advantage of our "good luck". Others in the same situation missed out, either because they hadn't considered what might happen, or they weren't ready when it did.

I've always been fascinated by the future, and the thought that our actions now can change what might happen later. This naturally led to an interest in risk management—proactively addressing uncertainty in a way that makes it more likely that we'll be successful. My first few years as a risk specialist were focused on preventing things from going wrong, or at least making them less likely and/or less severe. But I always felt there was something missing. Everything was focused on the negative, and the best outcome from risk management was that nothing bad happened.

Everything changed when I realised that not all uncertainty is bad. The future holds positive possibilities as well as potential problems. And if we can be proactive in addressing downside uncertainty, we can surely do the same for possible upsides. With that realisation, my role as a risk specialist took on a completely different complexion. I still had to help people prevent bad things, or make them less likely or less bad. But I could also help people find potential good things and capture them, or at least make them more likely or better. The idea of including opportunities within the scope of risk management, alongside threats, was born.

My first book shared my insights into this transformed way of understanding and managing risk. *Effective opportunity management for projects: Exploiting positive risk*[*] was published in 2003, explaining how to extend the traditional threat-focused risk process to address opportunities. This book broke new ground and was well received by businesses in various industries and countries, as well as being adopted by some universities as a textbook to support courses in project management or risk management. Some fellow risk practitioners were unsure about whether this was a development in the right direction, but in time a broad consensus appeared that the concept of risk includes both upside and downside.

Since then I've looked for others to take up the baton, waiting for books that would move things forward and add to my early insights. There have been surprisingly few, and my 2003 book has remained almost unique in addressing the challenge of managing opportunities in projects. Much has changed in the past 15 years or so, and we've gained a lot of useful experience in this area. Hence this book. My aim is to provide updated guidelines on how to apply opportunity management in the project arena, as part of an integrated risk process that addresses all forms of uncertainty. The earlier book was quite conceptual, focusing on ideas and principles as well as process. This book is more practical, with techniques, hints, and tips, offering real advice on how you can find and capture opportunities in your project.

---

[*] (Hillson, 2003)

If you know that current standards and guidelines for managing risk in projects recommend an inclusive approach, but you've wondered how to put this into practice in your own projects, this book is for you. If you feel that there must be more to risk management than "prevent and protect", you'll discover the answer in these pages. And if you're new to project risk management and want to make sure that you're following best practice right from the start, read on!

Dr David Hillson, The Risk Doctor<br>
Petersfield, Hampshire, UK<br>
April 2019

# About the Author

Known globally as *The Risk Doctor*, **Dr David Hillson** is a thought-leader and expert practitioner who speaks and writes widely on risk management. David has advised leaders and organisations in over 50 countries around the world on how to create value from risk, based on a mature approach to risk management, and his wisdom and insights are in high demand.

David has a well-deserved reputation as an excellent speaker and presenter on risk. His talks blend thought-leadership with practical application, presented in an accessible style that combines clarity with humour, guided by his motto: *"Understand profoundly so you can explain simply"*. He has also written 11 major books on risk, and over 100 professional papers. He has developed significant innovations that are now widely accepted as best practice, including broadening the risk process to cover proactive opportunity management.

David has received many awards in recognition of his ground-breaking work in risk management. For example, he was named inaugural *Risk Personality of the Year* in 2010–2011 by the Institute of Risk Management (IRM). This award honours the individual who has made a significant global contribution to improving risk management knowledge, whose experience and expertise enables them to effectively promote the message of strong risk management, and who provides inspiration to their peers in the risk profession. David has also been recognised as a *Chartered Fellow* of IRM.

David Hillson is an active member of the Project Management Institute (PMI®), where he is recognised as a risk thought-leader, and he has received multiple awards including the *PMI Fellow Award* (2010). He was Vice Chair for *The PMBOK® Guide—Sixth Edition* update project, and specifically responsible for the risk chapter. David was on the Steering Group for the update of the PMI Risk Management Professional (PMI-RMP)® certification, and he was a core author for the PMI *Practice Standard for Project Risk Management*.

David became an *Honorary Fellow* of the UK Association for Project Management (APM) in 2008, marking his contribution to developing the discipline of project risk management. He is a past Chairman of the APM Specific Interest Group (SIG) on Risk Management, and he has co-edited three major risk guides for APM.

David was elected a *Fellow of the Royal Society of Arts* (RSA) in 2007 to contribute to its Risk Commission. He led the RSA Fellows project on societal attitudes to failure, and is editor of *The Failure Files*. He is also a *Chartered Fellow* with the Chartered Management Institute (CMI), reflecting his broad interest in topics beyond his own speciality of risk management.

# Section A
# Why Opportunities Matter

# Chapter 1
# What Are Opportunities?

Risk management manages risk. The central idea behind this book is that risk is a broad concept that includes both opportunity and threat, and consequently risk management should also manage opportunities as well as threats. Starting from first principles, this chapter details the logical thinking that leads to a broader definition of risk and explains how we have come to the position today at which an inclusive approach to risk management is widely accepted and practised. The book's title, *"Capturing upside risk: Finding and managing opportunities in projects",* makes it clear that our main focus will be the application of this type of risk management to the world of projects, although the ideas are equally applicable to all other forms of risk management. As a result, Chapters 2 and 3 concentrate on where we might find upside risks in our projects and how we might go about managing them.

## FIRST PRINCIPLES

The best way to make the case for including opportunity in the definition of risk is through a set of self-evident axiomatic principles, starting with the following three generic statements:

- Uncertainty is everywhere.
- Not all uncertainty matters.
- Not all uncertainties that matter are bad.

## Uncertainty is everywhere

Simple observation is enough to persuade us (if we needed it) that life is uncertain. We can see this at large scale in areas such as global climate, international relations, science and technology, health, etc. It is also evident in business, with market volatility, policy changes, mergers and acquisitions, etc. Closer to home we find uncertainty in human relationships, including communities, teams, clubs, and families.

We tend to think of earlier times as being simpler and kinder, with our ancestors having fewer choices and more certainty. But history teaches us that human life has always been subject to uncertainty, often with unpleasant consequences such as revolution, plague, war, or famine, but sometimes with good outcomes such as the Renaissance, discovery of antibiotics, or democracy.

One result of the pervasively uncertain context within which we live has been the development of coping mechanisms, which seek to address aspects of uncertainty and impose control on apparent randomness. We might view human culture, law, science, art, religion, or philosophy as attempts to find order in chaos. It seems that many (most?) of us are reluctant to accept our inability to understand or manage all uncertainty, perhaps driven by the potential that uncertainty has to generate unexpected consequences for which we have not planned or prepared.

## Not all uncertainty matters

If we tried to count the number of uncertainties in the universe, we'd soon give up. Uncertainty seems to be limitless. But despite its universal nature, we don't need to concern ourselves with each and every uncertainty that's out there. Fortunately, most of the uncertainties in the universe don't matter to you or to me. We can safely ignore any uncertainty that would not affect us in any way, whether it happened or not, whether it was true or imagined, whether it was large or small.

This reassuring fact provides us with a way of approaching the uncertain universe, offering a filter that allows us to discount the vast majority of uncertainties because they don't matter to us. There are a limited number of uncertainties that we need to know about, think about, talk about, and prepare for, and those are the uncertainties that matter.

But how do we know which those uncertainties are? What filter can we use to discard irrelevant uncertainty and leave only uncertainty that matters?

We define what matters through our objectives. These are the things that we are trying to achieve, by which we measure success or progress. Objectives exist at all levels of human endeavour. At the global scale we have objectives on big issues such as climate change, poverty reduction, child safety, adult literacy, health improvement, and so on. Businesses are built around a hierarchy of objectives, from strategic corporate objectives, through more detailed programme or functional objectives, and down to tactical project and operational objectives. Communities, families, and individuals also set objectives to define the things that matter to them, including relationships, health, career, quality of life, etc.

This focus on objectives allows us to refine our area of interest when we are considering uncertainty. Personally, I only need to be aware of uncertainties that matter to one or more of my personal objectives, and then seek to deal with just those. Professionally, my organisation needs to know about any uncertainty that could affect achievement of its strategic, functional, or tactical objectives. Politically, governments need to consider uncertainties that affect the wellbeing, security, and prosperity of the nation. The same is true of any level of human activity: The only uncertainties that need our attention and action are the uncertainties that matter, and what matters is defined by our objectives.

## Not all uncertainties that matter are bad

Clearly, we must be concerned about any uncertainty that has the potential to affect achievement of our objectives negatively. Where there is the possibility of loss, delay, overspend, injury or death, damage to reputation, reduction in market share, fall in share price, reduced competitiveness, or diminished stakeholder perception, we need to

- Understand the source and extent of the uncertainty
- Be clear about how and why it might affect us, and how badly
- Find ways to protect ourselves from the uncertainty and/or its adverse effects.

However, if we become aware of an uncertainty that would result in a positive outcome, assisting us to achieve our objectives, then this too would deserve our attention and action. Uncertainties that might produce savings, reduce timelines, improve safety records, enhance reputation, raise market share or share price, increase competitiveness, or heighten stakeholder perception are also important. It would be remiss of us to ignore such uncertainties; instead we need to ensure that we understand them fully and take whatever action we can to obtain the benefits that are on offer, maximising our chances of achieving our objectives.

So, although it is true that the world is full of uncertainty and that a subset of those uncertainties matter (the ones that could affect our objectives), it is also undoubtedly the case that uncertainties that matter include both bad things and good things, and we should be interested in both.

## RISK IS "UNCERTAINTY THAT MATTERS"

We've established that we all need to be concerned about uncertainties that matter, and that these include both bad and good things. Over a period of time, the discipline of risk management has been developed to help us find and manage these uncertainties, based on the idea that **risk is "uncertainty that matters"**.

Although this short phrase is not a formal definition of risk, it forms the basis for most of the official definitions contained in international standards and guidelines, as shown in Table 1-1. The *uncertainty* dimension is acknowledged in these definitions either explicitly or through use of terms such as "probability," "potential," or "possible". The requirement for a risk to *matter* is described in terms of "achieving goals" or "consequences/effect/impact on objectives". The concept of risk as uncertainty that matters can be used as a test for whether something is a real risk: If it is not uncertain, then it is not a risk; and if it does not matter, then it is not a risk.

- All risks are *uncertain,* although not all uncertainties are risks. There may be uncertainty around whether a risk will occur or not, and if so, when, or what might trigger its occurrence. Or we may be unsure of the potential extent of its effect on objectives, or how easy it might be to manage it.
- All risks *matter.* If they occur, risks will have consequences that make a difference in some way. By definition, it is not possible to have an inconsequential risk, because risks are inextricably linked to objectives.

One of the key principles discussed above is that not all uncertainties that matter are bad. If risk is "uncertainty that matters", then it automatically follows that not all risks are bad. This is recognised by most of the international standards and guidelines listed in Table 1-1, which mention "positive or negative impacts" or use the terms "opportunity and threat".

A threat is uncertain because it may never happen, and it matters because if it happened it would make it harder for us to achieve one or more of our objectives. So, a threat is an uncertainty that matters; in other words, it is a risk.

Similarly, an opportunity is uncertain because we have no guarantee that it will occur, and it matters because its occurrence would help us to achieve one or more objectives. An opportunity is therefore also an uncertainty that matters, and so it is also a risk.

## Table 1-1 Definitions of Risk as "Uncertainty That Matters"

| Source of Definition | "Uncertainty . . ." | ". . . That Matters" |
|---|---|---|
| *A Guide to the Project Management Body of Knowledge (PMBOK® Guide)*, Sixth Edition. (Project Management Institute, 2017) | "An *uncertain* event or condition . . ." | ". . . that, if it occurs, has a positive or negative *effect on* one or more project *objectives*." |
| *APM Body of Knowledge*, Sixth Edition. (Association for Project Management, 2012) | "An *uncertain* event or set of circumstances . . ." | ". . . that would, if it occurred, have an *effect on* achievement of one or more *objectives*." |
| | "The *potential* of a situation or event . . ." | ". . . to *impact on* achievement of specific *objectives*." |
| ISO31000:2018 *Risk Management Guidelines*. (International Organization for Standardization, 2018) | "Effect of *uncertainty* . . ." "An effect is a deviation from the expected. It can be positive, negative, or both, and can address, create, or result in opportunities and threats." | ". . . on *objectives*." |
| *Management of Risk [M_o_R]: Guidance for Practitioners*, Third Edition. (Office of Government Commerce, 2010) | "An *uncertain* event or set of events . . ." | ". . . that, should it occur, will have an *effect on the* achievement of *objectives*." |
| | "A risk is measured by a combination of the *probability* of a perceived threat or opportunity occurring . . ." | ". . . and the magnitude of its *impact on objectives*." |
| *PM² Project Management Methodology Guide*, Open Edition. (European Commission Centre of Excellence in Project Management, CoEPM², 2016) | "An *uncertain* event or set of events (positive or negative) . . ." | ". . . that, should it occur, will have an *effect on the* achievement of project *objectives*." |
| *Risk Analysis and Management for Projects: A strategic framework for managing project risk and its financial implications*, Third Edition. (Institution of Civil Engineers and Institute & Faculty of Actuaries, 2014) | "A *possible* occurrence . . ." | ". . . which could *affect* (positively or negatively) the achievement of the *objectives* for the investment." |
| *Standard for Risk Management in Portfolios, Programs and Projects*. (Project Management Institute, 2019) | "An *uncertain* event or condition . . ." | ". . . that, if it occurs, has a positive or negative *impact* on one or more enterprise, portfolio, program, and project *objectives*." |

The inevitable conclusion from this logical train of thought is that the concept of risk includes both upside risk (opportunity) and downside risk (threat), both of which matter, and both of which need to be managed proactively.

## THE DEFINITION DEBATE

The definition of risk has been a matter of debate for centuries, and the meaning of this term has changed with time more than once, with professionals often interpreting it in a way that was different from general usage.

Early definitions of risk included both upside and downside, going back as far as 3000 years ago with the Chinese character for risk, *wei ji*, which has two elements: *wei* means "danger", *ji* means "opportunity", together they mean "risk".

Fast forward to the Middle Ages, when the early Italian word *risicare* (or its alternative form *rischiare*) was translated as "to dare" or "to take the plunge". It had a particular use in merchant shipping, when the ship was "risked" in order to undertake a voyage which could end in wealth or disaster. The marine insurance market developed to cover the "risk", using premiums to provide a payout in case the ship was lost. Over time, the idea of insuring the *risk* became synonymous with insuring the *loss,* and the word "risk" began to take on a negative association.

This negative connotation persisted and strengthened until modern times, when the common perception is that risk is always and only bad. However, we do still recognise the potential for upside uncertainty—for example, when we hear statements such as, "*The value of your investment may go down as well as up.*"

A step change occurred in the way risk was used in a professional setting around the start of the 21st century. There was an increasing feeling that risk management was setting the wrong tone for businesses and their projects by focusing exclusively on what could go wrong and how bad it might be. This was demotivating to teams and frustrating for management, especially when additional funds and resources were requested to address risks that might never happen, or when large contingency amounts were set aside to cover risks that didn't occur. This led some risk practitioners to rediscover the lost idea that risk was a double-sided concept, including both upside and downside. A debate began over whether "risk" and "threat" were synonymous, or whether "risk" should properly be understood as encompassing both "threat" and "opportunity".*

After considerable discussion, some of it quite heated on both sides, the definition debate seemed to be resolved in the early years of this century, when international standards and guidelines began to use the broader definition of risk, and leading organisations began to implement an approach to managing

---

* (Hulett, Hillson, & Kohl, 2002)

risk that aimed to minimise threats as well as to maximise opportunities. As Table 1-1 shows, this position is now reflected by nearly every major risk management standard, particularly those that address the world of projects.

## THE GREAT DIVIDE

Despite the apparent conclusion of the definition debate in favour of the broader position, there remains a significant divide between theory and practice. Although the standards and guidelines are almost unanimous in declaring that risk includes both threat and opportunity, and that risk management should address both equally, many organisations still persist in limiting their risk management approach only to threats. Some recognise the existence of opportunities but treat them separately from threats, while others have no structured way for dealing with opportunity at all.

In 2001, we conducted a short survey to find out how organisations understood and managed risk. This survey also explored the views of risk practitioners to see whether they differed from the approach adopted by their organisations. The results were presented in detail in an earlier book (*Effective opportunity management for projects*).* This survey was repeated in 2019 to see how perception and practice had changed in the intervening years, asking the same four questions from the original survey, shown in Table 1-2. Sample size was higher for the later survey, owing to changes in communication technology—with 186 responses in 2001 and 473 responses in 2019—but the differences are still instructive.

The overall results for each question from the two surveys are shown in Figures 1-1 to 1-4, comparing responses from 2001 with 2019. More detailed analysis is possible, looking at differing perceptions across age groups, gender, geography, or industry type, but the headline results are sufficient to draw some important conclusions.

The first two survey questions explored the risk approach adopted by the organisations represented by the respondents. Q1 asked how organisations define risk, and Q2 looked at the type of risk process they employed.

- *Definition of risk* (Figure 1-1). In 2001, over half of the respondents (54%) reported that their organisations used a negative definition of risk, exclusively associating risk with threats, and about a third (34%) used a broader definition that included both threats and opportunities. In 2019 these

---

* (Hillson, 2003)

**Table 1-2 Risk Definition Survey Questions**

| Survey Question | Response Options |
| --- | --- |
| Q1. Which of the following definitions of risk is closest to that used by your organisation? | (a) Neutral definition, for example: "An uncertain event or condition which, if it occurs, would have an undefined or unknown impact on achievement of objectives."<br>(b) Negative definition (threat), for example: "An uncertain event or condition which, if it occurs, would have a negative impact on achievement of objectives."<br>(c) Inclusive definition (threat & opportunity), for example: "Risk is an uncertain event or condition which, if it occurs, would have a negative or positive impact on achievement of objectives."<br>(d) Other |
| Q2. Which of the following best describes your organisation's approach to risk management? | (a) The risk management process aims to manage potential negative impacts on objectives (i.e., threats only). There is no process for explicit handling of opportunities.<br>(b) The risk management process aims to manage potential negative impacts on objectives (i.e., threats only). Opportunities are handled via a separate process that is not an integrated part of risk management.<br>(c) The risk management process aims to manage both threats and opportunities in a common (integrated) process.<br>(d) Other |
| Q3. Which of the following definitions of risk best reflect your own preferred definition of risk? (Note: This may be different from the definition used by your organisation—see Q1.) | (a) Neutral definition, for example: "An uncertain event or condition which, if it occurs, would have an undefined or unknown impact on achievement of objectives."<br>(b) Negative definition (threat), for example: "An uncertain event or condition which, if it occurs, would have a negative impact on achievement of objectives."<br>(c) Inclusive definition (threat & opportunity), for example: "Risk is an uncertain event or condition which, if it occurs, would have a negative or positive impact on achievement of objectives."<br>(d) Other |
| Q4. Do you support the recent change by standards organisations and professional bodies to define risk as including both threats and opportunities? | (a) Yes<br>(b) No<br>(c) Don't know<br>(d) Don't care |

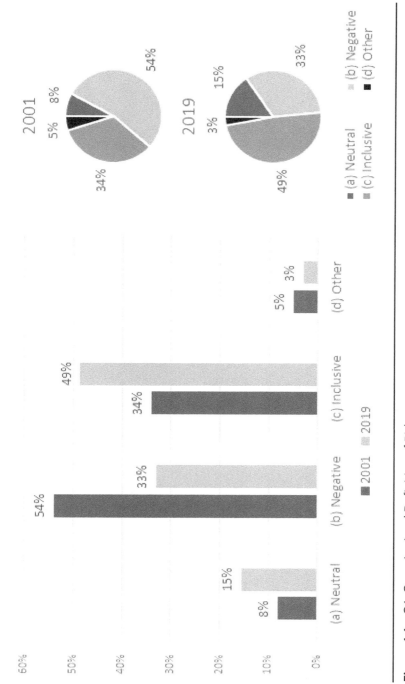

**Figure 1-1** Q1. Organisational Definition of Risk

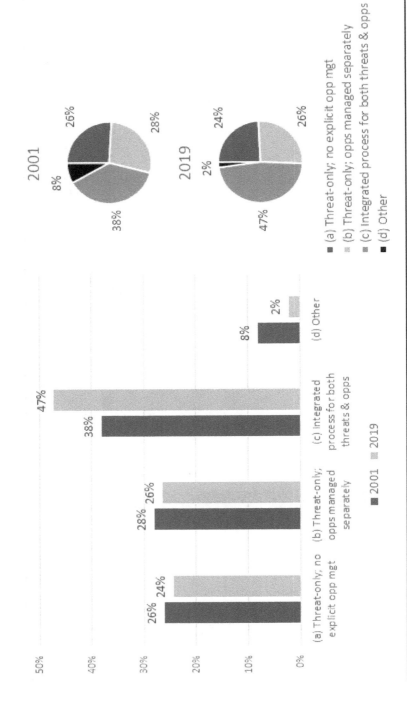

**Figure 1-2** Q2. Organisational Approach to Risk Management

proportions were almost exactly reversed, with 33% of organisations defining risk as only negative, and 49% taking an inclusive approach. The number of organisations whose definition of risk was neutral increased between 2001 (8%) to 2019 (15%).
- *Focus of risk process* (Figure 1-2). The survey question offered two options in which the risk process only addressed threats. In one of these options there was no process for dealing with opportunities, and in the other the organisation managed opportunities separately from threats. Combining these two options gives the total proportion of organisations whose risk process is threat-focused. In 2001, this represented 54% of organisations, and the 2019 results showed a slight decrease to 50%. Responses also indicate that the proportion of organisations using an integrated risk approach that covered both threats and opportunities has increased slightly from 38% in 2001 to 47% in 2019.
- Taken together, these results reveal an interesting consistency between policy and practice, which appears to have persisted between 2001 and 2019.
  o In 2001, there was a close correlation between the risk definition and the focus of the risk process, in which 54% of organisations used a negative definition, and the same proportion (54%) had a threat-only risk process. The proportion with an inclusive definition of risk was 34%, and 38% had an inclusive risk process.
  o In 2019, the correlation holds for organisations with an inclusive definition of risk, the majority of which (77%) also have an inclusive risk process, and for organisations defining risk as negative, of which 88% also have a threat-focused risk process.

The last two questions focused on the views of individual respondents. Q3 in the survey asked about their personal preferred definition of risk, and Q4 sought their view on the change to inclusive risk definitions found in most international standards and guidelines.

- *Individual preferred definition of risk* (Figure 1-3). The results have changed significantly between 2001 and 2019, with fewer respondents seeing risk as exclusively negative (decreased from 33% to 11%), and more adopting the inclusive definition (up from 46% to 74%).
- *Support for definition changes in standards* (Figure 1-4). In the 2019 results, 84% of respondents support the move by standards and guidelines to define risk more broadly (up from 60% in 2001), with only 9% opposed (down from 30% in 2001). Interestingly, comparing 2019 results from Q3 and Q4, 95% of those who see risk as inclusive say they support the

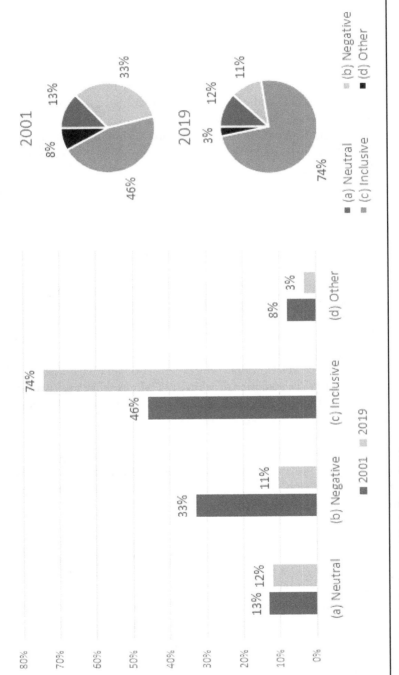

**Figure 1-3** Q3. Personal Definition of Risk

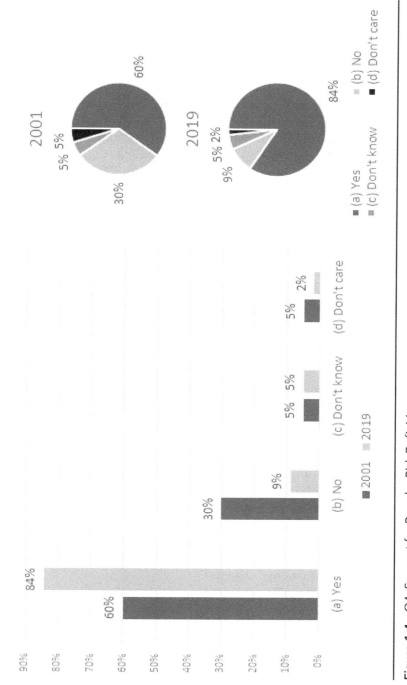

**Figure 1-4** Q4. Support for Broader Risk Definition

change in standards (as expected), but only 46% of the individuals with a threat-only definition of risk oppose the change.
- *Individuals and organisations* (Figures 1-1 and 1-3). Comparing Q1 and Q3 also allows us to test whether the way individual respondents view risk might differ from the definitions used by their organisations. In 2001, the proportion of *individuals* defining risk as including both threat and opportunity was significantly higher than the proportion of *organisations* taking that view (46% compared to 34%). This is also shown in the 2019 results, but the difference is more pronounced (74% for individuals, compared to 49% for organisations). Similarly, fewer individuals took a threat-only position than did their organisations in 2001 (33% vs. 54%), and the same is true in 2019 (11% vs. 33%).

What do these results tell us about changes in the perceptions of risk and the practice of risk management since 2001? Overall, the changes are encouraging for those who see risk as a broad concept covering both threat and opportunity, and those who believe that the risk process should manage both. There are clear increases in the number of organisations and individuals who adopt the more inclusive approach, and matching decreases in those who focus only on threats. Changes in international standards and guidelines to define risk more broadly are overwhelmingly supported, with only a few disagreeing.

These positive moves suggest that we are closer to a settled position that risk includes both threat and opportunity, and risk management should address both threats and opportunities in an integrated process. However, the results are not entirely rosy, and it appears that there is still work to be done before this approach is universally accepted and adopted.

Although more organisations have taken on board the inclusive definition of risk than was the case in 2001 (increasing from 34% to 49%), and the proportion of organisations using an inclusive risk process has gone up similarly (from 38% to 47%), still a third of organisations view risk as exclusively negative, and half have a threat-focused risk process. It seems that recent changes in international standards and guidelines haven't convinced everyone that this is the right path to follow. The specific battle over how risk should be formally defined has been won, but the overall war is still underway in terms of persuading organisations to implement the required changes.

## REASONS FOR RELUCTANCE

International standards and guidelines agree that risk includes both threat and opportunity and that risk management should address both in a single risk process, but this is still not universally adopted in practice. What lies behind this hesitancy to adopt a broader approach?

Risk practitioners continue to ponder this question, conducting research in an attempt to uncover the reasons for reluctance. Although there are doubtless many causes with varying degrees of influence, the following five factors are important contributors to the lack of take-up:

- Ignorance
- Language
- Culture
- Psychology
- Inertia

## Ignorance

Those of us who routinely include opportunities within the risk process are constantly surprised by the number of times people claim they've never heard of such a thing. Although we shouldn't expect busy leaders and managers to be fully aware of the latest developments in international standards and guidelines, still the idea that risk includes both opportunities and threats has been around for so long that it seems reasonable to expect most people to at least be aware of it as a possibility.

Regardless of the underlying reason, ignorance is a frequent barrier to adoption of a wider risk approach that covers upside and downside risk equally. Most people are familiar with the legal principle that ignorance of the law is no excuse ("*ignorantia juris non excusat*"). This rule prevents individuals from avoiding prosecution by claiming that they did not know their conduct was illegal. But ignorance can, under certain limited circumstances, provide a viable defence to a criminal charge. This includes lack of sufficient public notice that something is a crime, and crimes that require wilful disregard. If a defendant can show that there were insufficient communications to make them aware of the law and that they therefore violated it unintentionally, they may have a valid defence, although there is a responsibility to make oneself aware of applicable laws where reasonably possible. Consciously avoiding knowledge of relevant laws is no defence!

In the (non-legal) case of the nature of risk, it seems unlikely that similar defences would apply, because risk management guidelines are publicly available, and professionals should be aware of them. Ignorance should be no defence! Nevertheless, perhaps risk management professionals and standards bodies could do better in communicating the message to businesses and practitioners.

## Language

A common objection heard by risk management trainers and presenters when they mention that risk includes opportunity as well as threat is that common-use language doesn't support this meaning. The lay person "knows" that risk is a bad thing, to be avoided or minimised wherever possible. The word risk itself "always" has negative connotations. This objection is also voiced in countries where languages other than English are spoken. It seems that risk means something negative in most languages.

We've briefly mentioned above that the concept of risk was double-sided in past times, and a few people are persuaded by the argument that this broader meaning has been lost recently and could/should be recovered. But for the majority, the common-use meaning of risk as negative is a significant barrier to accepting the inclusion of opportunities as part of the risk process.

A different response to this difficulty is to recognise that many words are understood by lay people in a way that differs from how they are used by professionals. Examples might include words such as schedule (a timetable for a bus or train, not an activity network showing logical dependencies), to estimate (guess the value of something, not provide a figure for budgeting purposes), buy-in (order a take-away meal, not gain acceptance or commitment), gateway (entrance to a property, not a transition point between two phases), etc. Phrases constructed from ordinary words can have special meanings, such as earned value, benefits management, or work breakdown structure. Following this line of thought, there's no problem if professionals mean something by "risk" which is different from what non-professionals understand. Every profession has its own jargon and special language, and we might argue that giving "risk" a double-sided meaning is just another example of business jargon.

Just as newcomers to a particular profession or discipline soon learn the jargon and become proficient speakers of the local language, so people who are new to the idea that risk means more than just threat can get used to it with time, and eventually find themselves "speaking risk" like a native!

## Culture

Culture is a hot topic, and a big one that we'll return to before the end of this book. But we need to mention culture here as one of the reasons for reluctance to implement a risk approach that includes opportunity. If we define culture as *"the values, beliefs, knowledge, and understanding shared by a group of people with a common purpose"*', we can see that culture exists at many different levels,

---

\* (Institute of Risk Management, 2012a, b)

depending on which "group of people" we're talking about, and each level of culture influences thinking and behaviour to a different degree.

Risk culture is a subset of general culture and is *"the values, beliefs, knowledge, and understanding **about risk** shared by a group of people with a common purpose"*\*. If the culture leads to risk being viewed as exclusively negative, then people's thinking and behaviour will naturally reflect that perception. We might consider this influence of culture at three levels:

- *National culture.* Pioneering work by Geert Hofstede in the 1980s[†] identified five distinguishing characteristics of national culture across a wide range of countries, one of which was Uncertainty Avoidance. This indicates how far members of a culture feel threatened by uncertain or unknown events and the degree to which people therefore seek to avoid uncertainty or ambiguity. Hofstede calculated an Uncertainty Avoidance Index (UAI) for 50 countries and three regions and found higher scores in Latin countries, in Japan, and in German-speaking countries, with lower scores in Anglo, Nordic, and Chinese culture countries. Hofstede proposed a series of national characteristics linked to UAI, summarised in Table 1-3.

**Table 1-3 National Characteristics Driven by Uncertainty Avoidance**

| Higher-UAI Countries | Lower-UAI Countries |
|---|---|
| Higher anxiety level | Lower anxiety level |
| Concerned about the future | Taking life a day at a time |
| Driven by fear of failure | Driven by hope of success |
| Committed to hierarchical structures | Prepared to bypass hierarchy where justified |
| Resisting change | Prepared to embrace change |
| Seeking consensus | Recognising the value of competition & conflict |

It's tempting to conclude that high UAI corresponds to risk-aversion, and low UAI represents risk-seeking, but Hofstede says this is an oversimplification: *"Uncertainty avoidance does not equal risk avoidance. . . . Uncertainty-avoiding cultures shun ambiguous situations. People in such cultures look for structure in their organisations, institutions, and relationships, which makes events clearly interpretable and predictable. Paradoxically, they are often prepared to engage in risky behaviour in order to reduce*

---

\* (Ibid.)
[†] (Hofstede, 1982)

*ambiguities. . . . Countries with weaker uncertainty avoidance tendencies demonstrate a lower sense of urgency. . . . In such countries not only familiar but unfamiliar risks are accepted."*

Nevertheless, the national characteristics in Table 1-3 will influence the way members of a society think about upside risk or opportunity, which will in turn affect their behaviour. People from higher-UAI countries are more anxious about all forms of risky behaviour, which could make them unwilling to take opportunities in case things go wrong in future. By contrast, lower-UAI cultures are less anxious and less concerned about the future, which might make them more willing to explore opportunities.

From the various levels of cultural influence, national culture exerts the weakest effect, but the underlying values and beliefs are deep-seated and hard to change.

- *Organisational culture.* The values and beliefs of an organisation are revealed in many ways, including communication style, remuneration policy, marketing stance, attitudes to staff, etc. In terms of risk, the tone from the top set by senior leaders will influence risk-related behaviour throughout the organisation, including decision-making at all levels. Where leaders adopt an entrepreneurial attitude combined with encouragement to explore and innovate in pursuit of progress or competitive edge, as well as tolerance of mistakes genuinely made and bad news honestly shared, then people across the organisation will feel more comfortable in taking risks. This will include a perceived freedom to explore opportunities that might offer significant advantages or savings. On the other hand, if the leadership tone is more "command and control", requiring strict adherence to procedures, frequent detailed progress reporting, and punishing failure and non-compliance, this is likely to inhibit people from pursuing opportunities and also make them more risk-averse towards threats.

    Organisations have recently become more aware of the importance of culture in general, and risk culture in particular, and many businesses are seeking to be more intentional in developing a strong and mature culture that supports and encourages an appropriate level of risk-taking, including the proactive identification and management of both threats and opportunities. Organisational culture has a stronger influence on behaviour than wider societal or national norms, and it should therefore receive more attention and action from businesses that recognise the need to take the right risks safely.

- *Project team culture.* The strongest cultural influence in project-based organisations comes from the smallest group to which project staff belong, which is the team itself. While it is common for team culture to reflect organisational culture quite closely, this doesn't have to be the case. Project managers who understand the importance of culture can take proactive steps to shape a context for the project team that encourages the most productive behaviour. This includes simple things such as team briefings, meeting rules, work allocation, progress reporting, and so on. Where project team members feel empowered to act within clear boundaries of responsibility and accountability, supplied with the necessary resources to complete their tasks, they are more likely to be comfortable with identifying and pursuing opportunities as part of the project risk process.

## Psychology

While working on human motivation, Abraham Maslow* developed a "hierarchy of needs", which suggests that people are motivated by the drive to satisfy needs, but that not all needs are equal. Maslow's hierarchy of needs shows the most powerfully motivating needs at the base, and those with less power at the top. An example of Maslow's hierarchy is shown in Figure 1-5.

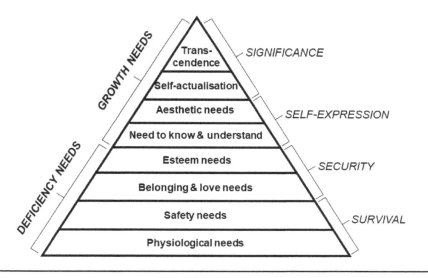

**Figure 1-5**  An Example of Maslow's "Hierarchy of Needs"

Maslow suggested that people are driven to satisfy lower needs before higher needs exert any influence. For example, the most basic physiological needs of air, water, sleep, and food must be met first, and these are the overriding concern

---

* (Maslow, 1943; 1987)

of each individual, even more important than being safe or feeling self-esteem. Once these needs are satisfied, a person is free to be concerned about other things. As each level of "hunger" is met (with literal physical hunger at the lowest level), higher needs emerge which require satisfying.

Maslow divided his hierarchy into two groups, with "deficiency needs" towards the base, and "growth needs" at the top.*

- *Deficiency needs* are those which must be satisfied, and without which a person feels something is lacking and might be said to be deficient or "needy". Deficiency needs are mostly physical and emotional, and when they are met, they cease to be active drivers of behaviour.
- *Growth needs* are those which add to a person, which are not necessarily required for a healthy existence, but which make a person more fully rounded and complete. They are psychological and spiritual, and they form more enduring and permanent motivators.

How is this relevant to the question of why individuals and organisations might find it difficult to implement opportunity management as part of an integrated risk process?

Threats are seen as deficiency needs, because they could jeopardise the health or even survival of our project or business. Opportunities appear in Maslow's hierarchy as growth needs, because they are perceived as optional, nice-to-have nonessentials. Maslow predicts that we will always be driven to address threats before opportunities, because the drive to survive is stronger than the attraction of growth. This suggests that basic human psychology leads us to focus first on threats and to treat opportunities as less important. In highly threatening situations, such as a very risky project or investment, we might even view opportunities as irrelevant distractions from the need to survive.

## Inertia

The final reason for reluctance to adopt opportunities as an integral part of the risk process is simple inertia. If the approach to managing risk in an organisation is threats-only, then people and projects will tend to follow current and previous practice, perhaps unthinkingly. Legacy processes and practices provide a straitjacket, constricting people's freedom of movement, and limiting their ability to consider upside risk. Indeed, in a strongly threat-focused environment, the very existence of opportunities may be beyond people's imagination. An upside risk becomes literally unthinkable.

---

* (Maslow, 1943; 1987)

Even where people or organisations are aware of the possibility of adopting a more inclusive risk process, it's often easier not to change, but simply to continue to do things "the way we've always done them", even if the old ways aren't really working and a better way is available.

## WINNING THE WAR

This chapter has shown that, despite the almost unanimous agreement of international standards and guidelines that risk includes both threat and opportunity, practice is still not keeping up with theory. Still a third of the organisations represented in the survey view risk and threat as synonymous, and half have risk processes that only address negative risks. The definition battle may have been won in the world of standards, but the outcome of the war remains to be settled in the field.

With so many reasons for reluctance, it seems a miracle that any organisations have ever decided to take up the proactive identification and management of opportunities as part of the risk management approach. Fortunately, for each of the most common reasons people give for not doing it, there are strong countermeasures to encourage them to take the plunge. These include:

- Education and demonstration (counters *ignorance*)
- Persistence (can change use of *language*)
- Memetics (modifies the influence of *culture*)
- Emotional intelligence (moderates *psychology*)
- Support (overcomes *inertia*)

Each of these countermeasures can encourage change on its own, but they are most powerful in combination.

### Education and demonstration

Ignorance is easily countered by education, and many good resources exist that explain the underlying thinking and practical benefits offered by a broader approach to managing both threats and opportunities. Formal training courses, webinars, online videos, websites, blogs, and books provide a rich source of knowledge and skills waiting to be tapped by individuals and organisations who recognise the need to update their approach to risk management.

Another important factor in persuading people to consider changing their approach is the ability to show how inclusive risk management works and why it helps. If we can demonstrate how simple it is to find and capture opportunities, and what benefits follow from doing so, others will want to follow our lead. We should record and share case studies in which opportunities were proactively identified and turned into concrete benefits. This can be done via conference presentations, journal and magazine articles, blogs, press releases, marketing materials, and websites.

## Persistence

Changing the way people talk about risk is a real challenge. The idea that all risk is bad is ingrained into our thinking, and using the word "risk" in a more inclusive way causes problems for most people, at least initially. Risk practitioners and professionals have struggled for decades to solve this language conundrum, without success. As we've seen above, standards organisations have decided to bite the bullet and redefine risk to have a broader meaning than the way laymen understand and use the term.

While it is undoubtedly difficult to change the way we speak about risk, it's certainly not impossible. If we persist in using the word risk to mean both threats and opportunities, and we do it consistently, eventually we'll develop new patterns of thought and speech. Experience shows that most teams adopt the new language, thought, and practice within a few weeks. Once this barrier is overcome, it is much easier to implement an inclusive approach to managing risk, without constantly tripping over the use of terms.

## Memetics

Richard Dawkins[*] proposed the existence of "memes" as self-replicating units of information (like a gene). A meme is an idea or concept which exists in the human mind and which replicates by transfer to other minds. The most successful memes are those which are best adapted to their environment, giving them a competitive advantage so they can replicate and become dominant ("survival of the fittest"). Culture is formed by the current group of dominant memes. Memetics describes how memes operate, and memetic engineering attempts to manipulate memes in order to produce a desired outcome. Memetics is entirely hypothetical, but it offers useful insights into many aspects of human behaviour and culture, including management of risk.

We can think of the concept of "all risk is bad" as a meme, and "risk includes both threat and opportunity" is another, with these two memes competing. Which meme is best adapted to its environment? Many aspects of our culture,

---

[*] (Dawkins, 1989)

including organisational culture, are risk-averse, uncomfortable with uncertainty and seeking to reduce it wherever possible. The "all risk is bad" meme seems better fitted to this environment, because it encourages more appropriate behaviour, including prevention, protection, contingency, mitigation, avoidance, etc. To a risk-averse person or organisation, including opportunity as a type of risk seems to increase risk exposure by suggesting that some risk should be embraced, exploited, or enhanced. In a risk-averse culture, the "all risk is bad" meme seems better adapted to survival and replication, whereas the "risk includes both threat and opportunity" meme is weaker and less competitive.

Memetic engineering suggests three approaches to make the "risk includes both threat and opportunity" meme more competitive:

1. *Modify the "risk includes both threat and opportunity" meme.* We can increase the survival value of a meme by making it more competitive than its rival in the current environment. The main strength of the "all risk is bad" meme is its apparent alignment with the prevailing social and business climate, based on fear of the unknown, the desire for predictability and control, and the view that uncertainty is bad and must be resisted. We can challenge this view by emphasising that proactive identification and capture of opportunities maximises the chances of success and protects against unforeseen or unexpected downside impacts.
2. *Actively promote the "risk includes both threat and opportunity" meme.* This strategy requires demonstrating its value in the current environment and/or maximising its perceived advantage in the expected future environment. We need public champions who will use the skills of marketing, advertising, and branding to communicate the benefits of proactively managing opportunities, using successful case studies wherever possible.
3. *Demonstrate the limitations of the "all risk is bad" meme.* We can also increase the competitive advantage of the "risk includes both threat and opportunity" meme by reducing the attractiveness of the "all risk is bad" meme. We should seek to persuade people that a limited focus on just downside risks will result in reduced benefits, lost value, degraded competitiveness, lack of innovation, poor creativity, etc. Demonstrating that the "all risk is bad" meme is an inadequate response to the current and future uncertain environment will reduce its ability to compete against the "risk includes both threat and opportunity" meme.

## Emotional intelligence

The way we think influences the way we act. If our mindset is threat-focused, especially in relation to our work and projects, then our behaviour towards risk will be aimed towards prevention and protection. If we think about risk in a broader way, recognising the possibility of upside risk as well as downside, then our risk-related actions will be correspondingly more inclusive.

We've seen above how Maslow's work on motivation explains why humans tend to focus first on threats. If Maslow is right, then it's entirely natural for us to adopt a threat-focused mindset. "Prevent and protect" is a deep-seated gut reaction in the face of uncertainty and risk. But people are not unthinking reactors, driven only by their instincts. We have the capacity for self-awareness and reflection, and the ability to moderate our behaviour to act in way that differs from our initial gut reaction. In recent years, the ideas of *emotional intelligence*[*] have been developed to describe this process of awareness, appreciation, assessment, and action. If our natural unmanaged way of thinking is preventing us from considering opportunities, we can (should?) exercise some emotional intelligence and change the way we think.

True emotional intelligence is not "fake it till you make it", pretending to act in a way that is contrary to our real underlying drivers, beliefs, or motivations. Instead, it involves a process of actually changing our internal environment so that we genuinely think and feel differently about something.

Where we encounter barriers among other stakeholders to adopting an inclusive approach to risk management instead of only considering threats, either because "we've always done it this way" or because "everyone knows risk is bad", we can lead others to adopt a more emotionally intelligent position, helping them to change their mindset in way that leads to a change in behaviour.

## Support

Many organisations have a risk management process that focuses only on threats. Fortunately, good work has been done in recent years to develop inclusive risk processes that cover threats and opportunities, giving equal attention to both. Examples include the latest bodies of knowledge from professional organisations in the project arena (see Table 1-1), as well as some proprietary methodologies (such as ATOM: Active Threat & Opportunity Management[†]). These inclusive risk processes also provide proven practical techniques for identifying, assessing, analysing, responding to and reporting on both threats and opportunities in an integrated and equitable way.

However, when it comes to tools to support a risk process and risk techniques that are inclusive, there is still work to do. Many risk management software packages either ignore the existence of opportunities or pay them lip service

---

[*] (Goleman, 1995)
[†] (Hillson & Simon, 2012)

without providing the full functionality that's offered for threats. We'll return to this issue in Chapter 13.

Nevertheless, the processes and techniques are in place to update legacy approaches to risk management, providing an answer to the inertia problem of, "We've always done it this way."

## FROM THEORY TO PRACTICE

Starting from first principles, we've seen in this chapter that it's entirely reasonable to consider risk as a double-sided concept that includes both threat (downside risk) and opportunity (upside risk). This position is now embodied in virtually all international standards and guidelines on risk management. A survey comparing perceptions of risk in 2001 with those in 2019 shows a significant uptake of this broader approach to managing risk, demonstrated by the definitions of risk used by many organisations and the risk management process they have adopted.

Some organisations seem more prepared than others to adapt their risk approach to include opportunities. In particular, three application areas are well suited to the pursuit of opportunities:

- Businesses with a *high level of innovation or R&D* are sensitive to the existence of upside risk and actively seek out areas of uncertainty that might be explored and exploited to create value, new products and services, or competitive advantage. Indeed, the innovation process itself can be seen as designed to find and capture opportunities that others might miss.
- *Entrepreneurs* also demonstrate an enhanced willingness to embrace uncertainty in search of opportunities. They may be particularly attuned to assessing and taking risk as a result of the personal stake that most entrepreneurs have in their businesses, especially in the early stages, and of course they will always be on the lookout for opportunities to help them move forward in ways that are faster, smarter, and cheaper.
- Finally, senior executives and leaders tend to become actively aware of opportunities at the strategic level when their organisations are *undergoing major upheaval or change,* usually resulting from external forces such as unforeseen marketplace developments, disruptive technologies, regulatory or political changes, and so on. These types of existential game-changer challenges focus the mind on the need to do things differently, either radically improving efficiency, developing completely new market offerings, restructuring the business or changing corporate strategy. The need to survive forces a change in management thinking and style, leading to a focused search for new opportunities.

There remain, however, a significant number of organisations whose thinking and practice both remain threat-focused. This chapter has outlined some reasons for reluctance, with possible remedies, and we'll return in the last chapter to consider what still needs to be done to persuade the hesitant to come on board. Before we get there, it's time to move from a general debate about the nature of opportunity and turn to one specific area of application in which opportunity management could be adopted more deeply: ***projects.*** The next two chapters explore where we might find opportunities in projects, and how we might go about managing them.

# Chapter 2

# What Are Opportunities in Projects?

The previous chapter explained why the concept of risk covers more than just threats, and made the case for a broader definition that also includes opportunities. Building on that insight, we can move on to consider how it might be applied to the world of projects. If the generic definition of risk includes both downside and upside, then it seems obvious that we should expect to find both threats and opportunities in our projects. And because risk management manages risk, it should be normal for project risk management to address both threats and opportunities in a common process. These two topics are dealt with in this chapter and the next.

## WHY PROJECTS ARE RISKY

Anyone who has ever worked on a project will know that projects are risky. But why? There are three main reasons:

1. Common characteristics
2. Deliberate design
3. External environment

## Common characteristics

All projects share a range of features which make them inherently uncertain:

- *Uniqueness.* Every project involves at least some elements that have not been done before, and the novelty and lack of experience with these elements introduces uncertainty.
- *Complexity.* Projects are complex in a variety of ways, and are more than a simple list of tasks to be performed. There are various kinds of complexity in projects, including technical, commercial, interfaces, or relational, each of which brings uncertainty into the project. A project can be viewed as a complex system, and if we are unable to predict the way the system will react, then uncertainty is inevitable.
- *Assumptions and constraints.* Project scoping involves making a range of guesses about the future, which usually include both assumptions (things we think will or will not happen) and constraints (things we are told to do or not do). Assumptions and constraints may turn out to be wrong, and it is also likely that some will remain hidden or undisclosed, so they are a source of uncertainty in most projects.
- *People.* All projects are performed by people, including project team members and management, clients and customers, suppliers and subcontractors. All of these individuals and groups are unpredictable to some extent and introduce uncertainty into the projects on which they work.
- *Stakeholders.* These are particular individuals and groups who can influence the project to a greater or lesser degree. Stakeholder interests can be varying, overlapping and sometimes conflicting, and the position adopted by stakeholders towards the project can change with time, making them a common source of uncertainty.
- *Change.* Every project is a change agent, moving from the known present into an unknown future, with all the uncertainty associated with such movement.

Any uncertainty arising from these characteristics that might affect project objectives should be considered as a risk. However, these characteristics are built into the nature of all projects and cannot be removed without changing the project. For example, a "project" which was not unique, had no constraints, involved no people, and did not introduce change would in fact not be a project at all. Trying to remove the risky elements from a project would turn it into something else, but it would not be a project.

We've seen that risk includes both threat and opportunity, so each of these sources of uncertainty might give rise to project opportunities. For example, *unique* aspects of a project might present unexpected savings in time and cost as a result of having to use novel development techniques. Considering *complexity* may reveal areas of synergy that were not previously apparent. Where *assumptions or constraints* limit the degrees of freedom for a project, if these were to prove false or flexible, it might allow the project to operate more efficiently. Not all *stakeholders* are adversaries of the project, and friendly stakeholders might offer support that makes life easier for the project team.

So, the fact that all projects share characteristics that make them inherently risky also leads us to expect that all projects will contain opportunities.

## Deliberate design

Any organisation seeking competitive advantage must recognise the relationship between risk and reward. Higher risk means potentially higher reward, though clearly there is also increased possibility of significant loss. By trying to make bigger changes more quickly, an organisation takes more risk in both dimensions—positive and negative. But the cautious organisation that takes less risk also reduces its potential for reward.

In project-based organisations, the role of projects is to deliver value-creating capabilities. As a result, projects are deliberately designed as risk-taking ventures. Their specific purpose is to produce maximum reward for the business while managing the associated risk. Because the existence of projects is so closely tied to reward, it is unsurprising that they are also intimately involved with risk. Organisations which understand this connection deliberately design their projects to take risk in order to deliver value. Indeed, projects are undertaken in order to gain benefits while taking the associated risks in a controlled manner.

Clearly, if we design our projects with risk in mind, we'll be looking to expose ourselves to as much upside risk as possible, positioning the project to take advantage of any opportunity that comes our way—as well as seeking to minimise and avoid exposure to threats. These aspects of managed and intentional risk exposure should form part of the design of the project.

## External environment

Projects are not conducted in a vacuum, but exist in an environment which poses a range of challenges and constraints. This includes both the wider organisational context and the environment outside the organisation, and changes beyond the project's control can occur in both of these. Environmental factors which introduce risk into projects include:

- Market volatility
- Competitor actions
- Emergent requirements
- Client organisational changes
- Internal organisational changes

Each of these factors is subject to change at an increasing rate in the modern world. Most projects have a fixed scope which they are required to deliver within this ever-changing environment. It is not possible to isolate projects from their environment, so this represents a common source of risk for projects. However, as for all risks, those arising from the external environment include both opportunities and threats. Markets can move in our favour, making our products and services more attractive. Competitor actions can also result in advantage to our project, perhaps allowing us to recruit skilled staff made redundant by others, or opening a new market that we can profitably exploit.

## SYNONYM CONFUSION

If projects are risky, as everyone agrees, and if the concept of risk includes both threat and opportunity, then every project should be able to find opportunities. Unfortunately, when we start to look for opportunities in projects, we immediately encounter a problem with language. We use the word "opportunity" in at least five different ways when it comes to projects, and only one of these relates to opportunity as upside risk. The four non-risk uses of the word opportunity in the context of projects are:

- The "opportunity to add or change something in the project", which generally involves a change in requirements or scope, either agreed or imposed.
- An "opportunity to act", which is a simple decision or choice to be made about a course of action (perhaps in response to a risk), not an uncertainty.
- An "opportunity to succeed", generally referring to a benefit or outcome that the project is intended to deliver.
- Our colleagues in portfolio or programme management roles might consider the project itself to be an "opportunity to deliver value".

None of these uses of the word opportunity relates to it as a risk.

In Chapter 1, we developed an initial definition of risk as "uncertainty that matters", and risks matter because they have the potential to affect our ability to achieve objectives. A project risk is "any uncertainty which, if it occurs, would affect achievement of one or more project objectives". Threats have unwelcome or adverse effects on achievement of objectives, and opportunities have helpful or beneficial effects.

*A project opportunity is therefore any uncertainty which, if it occurs, would have a positive effect on achievement of one or more project objectives.*

When seeking opportunities in projects, we must always remember that we are looking for *uncertainties*, not choices or facts or requirements or decisions. Project opportunities must also *matter*, by making it easier or faster or cheaper to achieve project objectives.

## LINKS BETWEEN OPPORTUNITIES AND THREATS

The definitions that we've looked at earlier make it clear that threats and opportunities are essentially the same thing: They are both uncertainties that matter. The only difference is in the sign of the impact: Threats have a negative impact, whereas opportunities have a positive impact.

This similarity might help us to understand the nature of opportunities in projects, by considering the relationship between threats and opportunities. There are at least four ways in which opportunities relate to threats:

1. Some opportunities arise from the *absence of threats*. If the bad thing does not happen, we might be able to take advantage of something good instead. If the threat is described by answering a "What if . . . ?" question, then we might find a corresponding opportunity by asking "What if not?" For example, if poor industrial relations do not lead to a strike, parties on both sides of the dispute might be incentivised to enter discussions aimed at turning the situation round from negative to positive.
2. Other opportunities are the *inverse of threats*. Where a variable exists on a continuous scale and there is uncertainty over the eventual outcome, instead of just defining the risk as the downside, it might also be possible to consider upside potential. Most variability is double-sided, with possible values both above and below what is planned or expected. For example, where the productivity rate on a new task is unknown, it might be lower than expected (a threat), or it might be higher (an opportunity).
3. We should also remember *secondary risks*, which are introduced by implementing a response to another risk. Secondary risks can be either threats or opportunities, just like any other risk. Sometimes by addressing one risk we can make things worse (the response creates a new threat), but it is also possible for our action to create a new opportunity. Avoiding potential delays to my car journey by taking the train might also allow me to do some useful work during the journey.

4. Lastly, we must not neglect "pure opportunities" which are *unrelated to threats*. These are simply unplanned good things which might happen. For example, a new design method might be released which we can apply to benefit our project. Or a new recruit to the team may unexpectedly possess a skill needed to solve a problem. This type of opportunity needs to be actively sought out, requiring fresh thinking and awareness of how potential additional benefits might arise.

## RISKS AND RISK

Before we leave the discussion of why all projects contain opportunities, there is one additional important topic to cover. This arises from the rather puzzling question, "What is the difference between risks and risk?"

When considering risk in projects, there are two levels of risk that should concern us.*

- The first level comprises the *individual risks within the project*, which are specific threats or opportunities that could affect our project's objectives. These are the risks that we record in our risk register, and they allow the project manager and team to answer the question, "What are the risks in my project?"
- Secondly, we need to understand the *overall level of risk exposure* associated with our project as a whole. When our project sponsor or client asks, "How risky is my project?", the answer does not usually come from the risk register. Instead of wanting to know about specific individual risks, we need to communicate the level of overall project risk. This represents the effect of all forms of uncertainty on the project as a whole, including individual risks, but also taking account of variability and ambiguity.

These two different perspectives reveal an important dichotomy in the nature of risk in projects. A project manager and the team are interested in "risks", while the project sponsor or client wants to know about "risk". While the project manager looks at *the risks **in** the project,* the project sponsor looks at *the riskiness **of** the project.*

This distinction is included in two of the main guidelines on project management, published by the Association for Project Management (APM) and the Project Management Institute (PMI®), respectively, as shown in Table 2-1.

Given these two levels of interest, any approach to risk management in projects needs to be able to answer the questions of both project manager and

---

* (Hillson, 2014b)

### What Are Opportunities in Projects?

**Table 2-1 Risks and Risk in Current Guidelines**

| Source of Definition | Individual Risks | Overall Project Risk |
| --- | --- | --- |
| *APM Body of Knowledge*, Sixth Edition. (Association for Project Management, 2012) | An uncertain event or set of circumstances that would, if it occurred, have an effect on achievement of one or more objectives. | Exposure of stakeholders to the consequences of variation in outcome, arising from an accumulation of individual risks together with other sources of uncertainty. |
| *A Guide to the Project Management Body of Knowledge (PMBOK® Guide)*, Sixth Edition. (Project Management Institute, 2017) | An uncertain event or condition that, if it occurs, has a positive or negative effect on one or more project objectives. | The effect of uncertainty on the project as a whole, arising from all sources of uncertainty including individual risks, representing the exposure of stakeholders to the implications of variations in project outcome, both positive and negative. |

project sponsor. An effective project risk management process should identify individual threats and opportunities within the project and enable them to be managed appropriately, and it should also provide an indication of overall project risk exposure.

However, when we are thinking about the place of opportunities in projects, "risks" and "risk" are rather different.

- There will be *many individual risks* in our project, including specific threats and opportunities, each of which can be identified, assessed, and prioritised, and for which we can develop and implement targeted risk responses. Throughout the various steps of the risk process, we *consider individual opportunities and individual threats separately.*
- There is only *one level of overall project risk exposure* for our project at any one time. Overall project risk fits the "uncertainty that matters" paradigm, with the "uncertainty" dimension expressing the degree of confidence that we have in achieving the project's overall objectives, and the "mattering" dimension expressed as the range of possible variation around each objective. In order to calculate these two dimensions of overall project risk, we need to take account of the combined effect of all individual threats and opportunities, as well as the impact of other sources of project uncertainty. The process of determining overall project risk exposure *does not consider threats and opportunities separately, but combines them together.* Chapter 7 describes how to calculate overall project risk in more detail.

## ALL PROJECTS INCLUDE OPPORTUNITIES

We've seen that all projects are inherently risky, and because risk includes both opportunity and threat, then it follows that all projects have opportunities. Unfortunately, loose language often causes confusion, with the word "opportunity" being used in several ways in the context of projects. If we want to extend our risk management process to include opportunities, we need to remember that we're looking for *uncertainties which, if they occur, would have a positive effect on achievement of one or more project objectives*. This definition gives us two key aspects that all opportunities in our projects share:

- *Uncertainty* . . . The first essential characteristic of all opportunities in our projects is that they are uncertain. They include uncertain future events that may never happen, as well as variability in planned activities that may turn out to be better than expected, and ambiguous or emergent aspects of the project where we may be pleasantly surprised.
- . . . *that matters*. The second crucial element of a project opportunity is that if it occurs it will be good for the project. Opportunities that happen will result in saved time, reduced cost, enhanced performance, reduced hassle, improved morale or teamwork, better client relations, etc. A realised opportunity doesn't change the project's objectives, but it makes it easier for us to achieve them.

Example opportunities from real projects include the following, each of which meets the two "uncertainty that matters" criteria:

- Because the client has expressed an urgent need for our product, we may be able to insist on experienced resources to serve on the project team, which would lead to improved quality, cost, and schedule performance.
- As a result of the good history of collaboration with our supply chain, interface management may experience fewer issues than expected, which would have a positive impact on the schedule.
- Because many other large projects are being executed concurrently, we might be able to negotiate better terms with the current fabrication subcontractor and capitalise on economies of scale, which would lead to cost savings and improved schedule.
- Because we have to outsource production, we may be able to learn new practices from our selected partner, which would lead to increased productivity and profitability.
- Because we have an existing product range, a higher level of technology reuse than planned may be possible, leading to savings in effort and cost.

- As the project is planned to take place during the summer, we may be able to recruit additional skilled student labour, which would mean that time could be saved on all activities that take place over that period.
- Because there are three other projects taking place in the same time frame, we may be able to utilise skilled staff as they become available from another project, which would allow us to deliver early to the customer.

In each case, the project team were made aware of something that might or might not happen, but that would be helpful to their project if it did occur. That opportunity awareness turned into proactive exploration and action, some of which resulted in significant positive impacts for the projects. These project teams were not super-professionals with access to some kind of hidden wisdom. Instead they simply followed a structured approach to finding and managing opportunities in their projects, and they reaped the benefits. But how?

Because opportunities and threats are both types of "uncertainty that matter" which only differ in the sign of their impact, we might expect to be able to use the same approach to identify, assess, prioritise, and respond to them both. It's time to consider process, and work out how we can effectively find and manage the opportunities that our projects undoubtedly contain.

# Chapter 3

# How to Manage Project Opportunities

The previous two chapters have explored the nature of risk in general, including the idea that it is a double-sided concept covering both opportunity and threat, and the fact that all projects are risky, meaning that both opportunities and threats are inherent, built in to every project. Knowing this, we can't just sit back and leave risk unmanaged on our projects. We need to understand the specific risks that we face, as well as the overall risk exposure of our project, and try to deal with them proactively, aiming to remove or reduce as many threats as possible, and capture or improve our opportunities. But how?

For a long time, many organisations either ignored the existence of opportunities within their projects, and so had no formal process for dealing with them, or they thought that opportunities were different from risks (threats), and so had a separate opportunity management process. We saw this in Chapter 1, where the results from an earlier survey conducted in 2001 showed 54% of organisations either having no explicit process for managing opportunities (26%) or addressing opportunities outside the risk management process (28%). The results from the 2019 survey were hardly different, with 50% of organisations in this position.

If risk includes both threat and opportunity, as we've seen, then risk management should manage both. We don't need a separate process. This chapter examines how and why.

## ELEMENTS OF A RISK MANAGEMENT PROCESS

The basis for saying that we can manage both threats and opportunities in a single integrated risk process is the idea that *an opportunity is the same as a threat*. The only difference is the sign of the impact: Threats have a negative effect on our ability to achieve project objectives, and opportunities have a positive effect. Otherwise, both threats and opportunities are "uncertainties that matter".

Once we see this similarity, the way to address opportunities becomes obvious. We can take the standard risk process which we already use for threats and apply it to opportunities, with simple modifications to recognise that we are dealing with positive upside risks.

Let's remind ourselves what a "standard risk process" should cover, whether it's dealing with just threats, or whether we're addressing both upside and downside risks. Different risk management guidelines and methodologies use a range of terms and titles for the various steps in the risk process, which can lead to confusion among users, especially those new to risk management. The best way to understand the components of risk management is to think about the underlying questions that each step aims to answer. Then risk management becomes a simple matter of asking and answering those questions.

### Nine structuring questions

Anyone responsible for managing a project will need to answer nine basic questions, and we can shape the risk management process around asking and answering them. The sequence of these questions is logical and natural, so the risk process based on the questions becomes intuitive and easy to follow, and less bureaucratic or forced. The nine questions are as follows:

1. *What are we trying to achieve?* We cannot start any project without first clearly defining its scope and clarifying the objectives that are at risk. We also need to know how much risk key stakeholders are prepared to accept, because this gives us the target threshold for risk exposure, and allows us to define levels of impact that will be regarded as high or low. We must address these factors as the first step of the risk process.
2. *What could affect us achieving objectives?* Once objectives and risk thresholds are agreed, we can start identifying risks, which are uncertainties that could affect achievement of objectives. There are a variety of risk identification techniques, each of which has strengths and weaknesses, so we should use more than one approach. In addition to considering individual risks, we should also identify sources of overall risk exposure which contribute to the riskiness of our project.

3. *Which of those things are most important?* Not all risks are equally important, so we need to filter and prioritise them. This will help us decide how to respond. When prioritising risks, we could use various characteristics, such as how likely they are to happen, what they might do to our objectives, how easily we can influence them, when they might happen, etc.
4. *How risky is the overall project?* In addition to prioritising individual risks, we should also consider the effect of overall risk exposure on the final outcome. This will need to take account of the individual risks, of course, but there are also other sources of uncertainty to be considered. Taken together, how do they affect the overall chances of project success, and what potential variation in project outcomes do they introduce?
5. *What shall we do?* Now we can start to think about what actions are appropriate to deal with individual risks, as well as considering how to tackle overall risk exposure. We might consider radical action, or attempt to influence the level of risk exposure, or decide to do nothing (apart from perhaps having a contingency plan). We might also involve other parties in responding appropriately to the risks.
6. *Did we do what we planned, and did it work?* We can plan to address risks, but nothing will change unless we actually do something. Planned responses must be implemented in order to tackle individual risks and change overall risk exposure, and the results of these responses should be monitored to ensure that they are having the desired effect. Our actions may also introduce new risks for us to address.
7. *Who shall we tell?* Various project stakeholders will want or need to know about specific individual risks or the overall level of risk exposure, and we need to communicate effectively with them. This means giving appropriate risk information to each stakeholder, delivered in the format and frequency that meets their needs.
8. *What has changed?* The risk process cannot end at this point, because risk is dynamic and changing. So, we have to look again at risk on a regular basis, to see whether existing risks have been managed as expected, and to discover new risks that now require our attention.
9. *What did we learn?* There is one more important step in the risk process, which is often forgotten. As responsible professionals we should take advantage of our experience with the past to benefit the future. This means we will spend time thinking about what worked well and what needs improvement, and recording our conclusions in a way that can be reused by ourselves and others.

In this series of questions, each one follows naturally from the one before. For example, once we are clear about our objectives, of course we'll want to know what might affect our ability to achieve them. We'll probably identify a long list of risks, too many to address them all with the same degree of effort, so of course we need to prioritise them. When we've discovered the most important risks, it is only natural to think about how we might address them proactively and effectively. And so on.

By structuring our risk process in this way, we make it easier for people to follow the process, as they will just be addressing a set of common-sense questions. Anything that makes risk management simpler will ensure that people are more engaged and that our risks are better managed.

## Tailoring the risk process

The question-and-answer approach to the risk process offers another key advantage, in addition to providing an intuitive framework for people to follow. The nine structuring questions also allow us to tailor the risk process to fit the particular risk challenge of our project. Clearly, we don't need to use a very detailed, complex process with multiple techniques if our project is small and simple. Conversely, we need more than a few informal team discussions to tackle risk adequately on a complex megaproject. Fortunately, the underlying risk process is the same for all types and sizes of project, based on the nine questions. All we have to do is answer the questions at a different level of detail.

For example, if you're managing a very small and simple project, you might choose to manage risk informally, perhaps even without any meetings or documentation. In this case, you could imagine driving to work or sitting on a train on a Monday morning, asking yourself the first five questions:

1. *What have I got to achieve today?*
2. *What might affect my ability to achieve that?*
3. *How do these things affect the overall riskiness of my project?*
4. *Which of those are most important for me to address?*
5. *What shall I do?*

Then when you get to work, you have a clear idea of your objectives (question 1), the main risks that you face (questions 2, 3, and 4), and your intended risk responses (question 5). While you're at work, you deal with the next two questions:

6. *Let's implement those responses—are they working as expected?*
7. *Who needs to know about these risks and my responses?*

On your journey home, you can reflect on the final two questions:

8. *What's changed today?*
9. *What have I learned?*

Then as you travel to work the next day, you start again with question 1. Alternatively, you may not need to do this every day, and you might ask the first set of questions on a Monday, spend the week managing your risks and communicating with others, then reflect on the week as you drive home (ready to relax over the weekend knowing that you've done your best to manage the risks you faced during the past week!).

The same nine questions can be used as a framework for a robust and fully detailed risk process for a megaproject, simply by answering the questions at a different level of detail. For example:

1. *What are we trying to achieve?* We might involve all our stakeholders in a series of requirements elicitation workshops, or undertake a value management exercise to define clear and agreed objectives. We can do this for the project as a whole, for defined subprojects, or perhaps only for the current phase of the project. This might involve incremental development of objectives, working collaboratively with our stakeholders, including clients and suppliers. We'll need to test that the set of objectives is coherent and internally consistent, that it is achievable, and that it fits with the strategic goals and direction of our organisation. On a large project, it may take several weeks or months before we have a fully agreed set of objectives.
2. *What could affect us achieving objectives?* As we consider the risks that we face on our megaproject, we'll want to use a variety of risk identification techniques, including learning lessons from past experience, examining carefully the characteristics and context of our current project, and thinking creatively about the future. We might set up a series of workshops involving subsets of our stakeholder community, each focusing on a different aspect of our project or current phase. We may use software tools to help us analyse the root causes of uncertainty, or to expose risks hidden within the structure of our project, perhaps including system dynamics modelling or futures thinking. This process of seeking out all the possible sources of uncertainty on our project could require considerable time and effort.
3. *Which of those things are most important?* Instead of just prioritising our risks on their probability of occurrence and size of potential impact, we might wish to use a number of other characteristics to determine which are most important. We might analyse connectivity between risks, to see if there are parent/child relationships with core risks affecting others.

4. *How risky is the overall project?* In considering the combined effect of all our risks and other sources of uncertainty, we might want to perform some deep analysis, quantifying risks and their dependencies, running what-if analyses, using a variety of techniques and comparing the results. It might be appropriate to look at overall project risk exposure through different lenses for different stakeholders, giving them a tailored perspective that matches their interest and role in the project.
5. *What shall we do?* When we start to decide on how to respond to individual risks and to the level of overall project risk exposure, it will be important to have a robust process for proposing alternative courses of action and choosing between them. We might use decision tree analysis, multi-criteria decision analysis, systems modelling, or real options to help us pick the most effective risk responses.
6. *Did we do what we planned, and did it work?* With each set of responses that we've decided to use, we should have defined success criteria and effectiveness measures, so that we can monitor how well our responses are doing, both in terms of addressing individual risk and changing the level of overall project riskiness. Rigorous identification of secondary risks will be necessary.
7. *Who shall we tell?* Risk communication on a megaproject is a major undertaking, and it needs significant attention and effort. We must first determine what risk information each stakeholder needs, why and when they need it, how to deliver it, etc. Then we need to create packages of risk information that will communicate clearly and unambiguously.
8. *What has changed?* Risk reviews on a megaproject will replicate the level of detailed analysis used for the initial identification and prioritisation of risks, involving multiple stakeholders, various techniques, working groups with a range of areas of interest, and so on. Each risk review might take a number of weeks to complete, in order to ensure that the understanding of risk exposure is comprehensive and current.
9. *What did we learn?* Megaprojects should have a fully formed and active process for identifying and learning lessons as the project proceeds. This will be integrated into an overall knowledge management process, and will include lessons relating to the risk management process. This allows us to learn as we go along, identifying and implementing lessons from earlier phases of our project to benefit later phases.

The risk process outlined here for a megaproject is very different from the process we described for the project manager looking after a small simple project. For one project, the risk process can be done informally and quickly, with a minimum of effort, but still covering all the necessary steps to identify, prioritise, and respond to the important risks. For the other, the risk process requires a high degree of formality and engagement, taking significant time and effort for the project team and other stakeholders. However, the risk process for both extremes of project type, from the simplest to the most complex, is based around *the same nine structuring questions*. The only difference is the level at which each question is addressed.

Once the underlying set of questions is understood and internalised by project managers and team members, risk management becomes easy to implement. Each question leads to the next, and the same process can be used on any project.

## USE A COMMON PROCESS FOR THREATS AND OPPORTUNITIES

We've seen how to use a series of questions as a framework to produce an intuitive risk management process, allowing us to work naturally and logically through the things we need to know and do in order to manage risk effectively. It also allows us to tailor the risk process to match the risk challenge of our particular project.

We've also seen that the concept of risk includes both opportunities and threats, and that an opportunity is the same as a threat, apart from the sign of the impact. This means that the generic "standard risk process" is not specific to managing threats; it works equally well for addressing opportunities.

### Nine structuring questions for opportunities

We showed above that using nine questions as a logical framework allows the risk process to be followed intuitively and naturally, since each question leads on to the next. How do these nine structuring questions apply to managing opportunities? Where might we need to modify the "standard risk process" to include opportunities alongside threats?

1. *What are we trying to achieve?* In this initial step, we clarify project objectives and define how much risk key stakeholders are prepared to accept. When setting risk impact levels, we also need to define what we mean by a high or low positive impact on project objectives for opportunities.
2. *What could **help** us to achieve objectives?* We can identify opportunities using the same techniques that work for threats. For example, we can hold a brainstorm session to think creatively about upside uncertainties, or we could produce an opportunity checklist based on previous

good experiences. Root-cause analysis or decision trees can help us find opportunities that might arise. And risk identification techniques such as SWOT Analysis or Force-Field Analysis naturally expose opportunities as well as threats.

3. *Which **opportunities** are most important?* As for threats, the importance of opportunities can be assessed in terms of probability ("How likely is the good thing to happen?") and impact ("How helpful would it be?"). Then we can use a standard prioritisation tool such as the Probability-Impact Matrix or a heat map to find the best opportunities.

4. *How risky is the overall project?* When we assess overall project risk exposure, it's really important to include both opportunities and threats in the calculation, otherwise we will inevitably produce a pessimistic result. Opportunity impacts are positive, producing savings in time or cost, or enhancing performance or reputation, etc., and they will improve the project's chances of success, as well as contributing to upside variation in outcomes.

5. *What shall we do about **opportunities**?* Having found some good opportunities that are worth pursuing, we can develop appropriate risk responses. This includes trying to exploit the best opportunities, and enhancing others to make them more attractive. We should also produce fallback plans to take advantage of any opportunities that might happen spontaneously. In the same way that threat responses aim to remove or reduce the negative effect of downside risks, opportunity responses are designed to capture or improve the positive effect of upside risks.

6. *Did we do what we planned, and did it work?* In the same way that we do for threats, we need to implement planned responses in order to tackle individual opportunities and improve overall risk exposure, and the results of these responses should be monitored to ensure that they are having the desired effect. Our actions may also introduce new risks for us to address.

7. *Who shall we tell?* Risk reporting must highlight the most important opportunities (alongside threats), informing stakeholders about where they need to pay attention and take action in order to maximise the potential upside.

8. *What has changed?* Risk reviews need to check whether existing opportunities have been managed as expected, and should seek additional opportunities that may have arisen since we last looked.

9. *What did we learn?* Risk-related lessons must include details of opportunities that we identified which might affect other similar projects, as well as opportunity responses that worked and that might be used again. This step of the risk process is particularly important for organisations that are new to managing opportunities through the risk process, so

that experience in identifying and capturing opportunities can be shared widely to benefit others.

It's clear that everything we know about downside risks (threats) is also true of upside risks (opportunities). Once we realise that an opportunity is the same as a threat apart from the sign of the impact, it will be easier to identify, assess, and respond to opportunities—we just use the same approach that already works for threats. And if we manage opportunities proactively, we will turn some of them into additional benefits, including reduced timescales, lower costs, or enhanced performance. This will result in more successful projects, which is good news for everyone.

## Benefits of integration

Because the same underlying nine questions can be used to identify, prioritise, and respond to important opportunities as well as applying to threats, it makes sense to use a single combined risk process to address both threats and opportunities together. There are a number of benefits that arise from using an integrated approach to managing risk, but perhaps the biggest three are as follows:

- *Effectiveness.* Everybody likes a bargain, and multi-buy offers in our supermarkets are very popular, such as two-for-the-price-of-one, or BOGOF (buy-one-get-one-free). By adopting an integrated risk process that addresses both threats and opportunities, we're getting double benefits from a single process. We can minimise threats and maximise opportunities in the same approach, and we don't need a separate opportunity management process.
- *Synergy.* As we've seen in Chapter 2, some opportunities have a direct relationship with threats, arising from the absence of a threat or the inverse of a threat. We can only take advantage of this synergy if we have an integrated common risk process that addresses both threats and opportunities together.
- *Serendipity.* This term comes from a Persian fairy-tale, "The three princes of Serendip", whose heroes were always accidentally discovering good and useful things that they weren't looking for. In the risk context it refers to the unexpected identification of opportunities, which is how many upside risks are first found, at least initially. The use of an integrated risk process encourages serendipity, because it requires us to think creatively about things that might never happen, about sources of uncertainty, and about areas in which we lack knowledge. We may initially set out to find and

manage downside risks. But as we start to imagine uncertain negative things that might hinder us from achieving our project objectives, it opens our thinking to begin to look for other types of uncertainty that might help us. In searching for downside risks, we might discover some upside risks as well.

## INTRODUCTION TO SECTION B

The first three chapters have made the case for including opportunity in the definition of risk, alongside threat, and explained how this is relevant to the world of projects. Finally, we've developed a structured risk process, based on asking and answering a series of logical questions, which allows us to address both threats and opportunities in a single integrated approach.

The next part of this book examines the risk process in more depth, focusing on what we need to do in order to find and manage opportunities in projects. This assumes that we'll use an integrated risk approach, which allows us to adapt some of our threat-focused risk techniques to deal with opportunities. The following chapters also use the nine structuring questions outlined in this chapter, although the chapter headings focus our attention on opportunities.

This nine-part risk process has many similarities with the way risk management is described in the main international risk standards and guidelines, as shown in Table 3-1. Some of these lack explicit steps for some of the questions, but most of them cover the same aspects of risk management, although perhaps in less detail. Table 3-1 also links the nine process steps to the chapters of Section B.

Table 3-1 Mapping the Nine Questions to International Risk Standards

| Structuring Question | Formal Process Step | A Guide to the Project Management Body of Knowledge, Sixth Edition. (Project Management Institute, 2017) | APM Body of Knowledge, Sixth Edition. (Association for Project Management, 2012) | ISO31000:2018 Risk Management Guidelines. (International Organization for Standardization, 2018) | Management of Risk [M_o_R]: Guidance for Practitioners, Third Edition. (Office of Government Commerce, 2010) | Mapping to Part B chapters |
|---|---|---|---|---|---|---|
| What are we trying to achieve? | Risk management planning | Plan risk management | Initiate | Scope, context, criteria | Identify—Context | Chapter 4: Setting the scene for opportunity management |
| What could affect us achieving objectives? | Risk identification | Identify risks | Identify | Risk identification | Identify—Identify the Risks | Chapter 5: Finding opportunities |
| Which of those things are most important? | Qualitative risk assessment | Perform qualitative risk analysis | Assess | Risk analysis risk evaluation | Assess—Estimate | Chapter 6: Picking winners |
| How risky is the overall project? | Quantitative risk analysis | Perform quantitative risk analysis | | | Assess—Evaluate | Chapter 7: Using numbers |
| What shall we do? | Risk response planning | Plan risk responses | Plan responses | Risk treatment | Plan | Chapter 8: Deciding what to do |
| Having taken action, did it work? | Risk response implementation | Implement risk responses | Implement responses | | Implement | Chapter 9: Taking action |
| Who shall we tell? | Risk communication | — | — | Communication and consultation/recording and reporting | Communicate | Chapter 10: Telling others |
| What has changed? | Risk review | Monitor risks | — | Monitoring and review | Embed and review | Chapter 11: Keeping up to date |
| What did we learn? | Identifying risk-related lessons | — | — | | | Chapter 12: Identifying risk-related lessons |

# Section B
# Managing Project Opportunities

# Chapter 4
# Setting the Scene for Opportunity Management

This chapter describes the first step in the risk management process. All projects are unique, with different levels of risk exposure, so we need to tailor the risk process to match the risk challenge faced by this project. That's why we include an initiation stage at the start of the risk process, deciding how to manage risk on this particular project.

If we're serious about identifying and managing both opportunities and threats in our project, we need to make our intentions clear right from the start. The tailoring decisions we make about how to manage risk on this project will include the fact that we expect to find and capture project opportunities and use them to improve our chances of success. In this chapter, we first describe the generic purpose and principles of risk management planning, then describe in more detail some of the more challenging aspects. At each stage, we'll explain where things are different for a risk process that covers both opportunities and threats. If we want to manage opportunities effectively in our project, we need to **set the scene**, so that everyone knows what to expect (and what is expected from them).

## PURPOSE AND PRINCIPLES OF RISK MANAGEMENT PLANNING

Most project teams are keen to identify risks. After all, we want to give ourselves the best chance of success, and that means knowing what risks are faced by the project so that we can get on and manage them, right? So, let's start now with risk identification, right?

Wrong! Immediately we try to start identifying and assessing risks, we come across a problem. Which techniques shall we use for risk identification? Who should we involve in identifying risks? What kind of risks are we looking for? How will we assess and prioritise them? If we intend to estimate how important each risk is, how do we know what "high probability" or "high impact" means? Who should we inform about the risks we've found, and how often will we need to review risks?

The answers to these and other questions will vary, depending on the nature of our project. The risk management approach we should adopt for a large and complex megaproject that makes a vital contribution to corporate strategy will be very different from the approach we might take for managing risk on a small simple project. This need for *scalability and tailoring* must be addressed before we start identifying, assessing, or responding to risks. Key factors to be considered include project size, project complexity, and the strategic importance of the project. This is why the risk process has a separate step before risk identification.

First, we need to plan our approach to managing risk on this project. Consequently, **the purpose of risk management planning is to ensure that the risk approach that we adopt for this project is appropriate and effective**.

A number of **principles** apply to this important first step in the risk process.

## Define objectives at risk and scope of risk process

The risk process cannot start until we know *why* we need to do risk management on this project. A key outcome from the risk management planning step is to define *what is at risk*, by clearly stating each of the main objectives of the project. These are often detailed elsewhere in the project documentation, such as the project charter or business case, but it is still important to state them clearly at the outset of the risk process. Risk is "uncertainty that matters", and it matters because if it happens it would affect achievement of project objectives. We also need to understand whether all project objectives are equally important, or whether one or more have higher priority than others. This will determine the way risks are prioritised, ranking a risk to a high-priority objective above a similar risk which might affect a lesser objective. We will need input on prioritising objectives from key project stakeholders, such as the project sponsor or business owner.

Having stated the project objectives, we also need to define the *scope of the risk process* that will be implemented. It is not always necessary to assess risk in relation to all project objectives, although this is most commonly the approach. In some circumstances the risk process may be focused only on one or two objectives, or on a particularly risky element of the project. Clear boundaries for the risk process must be set, so that the project team know what they are aiming to do.

## Reflect risk appetite of key stakeholders in measurable risk thresholds

"How much risk is too much risk?" This question allows us to understand the *risk appetite* of key stakeholders in relation to project objectives.* Risk appetite describes the underlying drive to take risk in order to gain reward, and it needs to be expressed in measurable *risk thresholds* against each objective. If we ask the project sponsor, "Is there any flexibility in the project end date or budget? How much tolerance does our client have for performance variation in final deliverables?", the answer is rarely "Zero" or "None", but we need to know *how much flexibility or variation* might be allowed. Without this information about risk thresholds, it is impossible to determine whether the level of risk in our project is acceptable or not. We can then use our understanding of risk appetite and risk thresholds to shape specific definitions of levels of probability and impact that will be used when prioritising risks.

## Tailor risk process to match the risk challenge of the project

Once we know what is at risk (objectives), the scope of the risk process, and how much risk is acceptable in the project, we can begin to tailor the risk process. If the purpose of risk management planning is *"to ensure that the risk approach that we adopt for this project is appropriate and effective"*, then we need to define what we mean by "appropriate and effective" for this project. This recognises that risk management is not one-size-fits-all. Although the basic steps of the risk process are the same in all cases, the level at which they are implemented on a particular project can and must vary.

As a result, the risk management planning step describes *how* risk management will be conducted on this particular project, including:

- Techniques to be used
- Reporting requirements
- Review and update frequency
- Risk-related roles and responsibilities

---

* (Hillson, 2012; Hillson & Murray-Webster, 2011, 2012)

## Summary

The purpose and principles of risk management planning are summarised in Table 4-1.

**Table 4-1 Purpose and Principles of Risk Management Planning**

| Purpose | To ensure that the risk approach on this project is appropriate & effective |
|---|---|
| Principles | • Define objectives at risk and scope of risk process<br>• Reflect stakeholder risk appetite in measurable risk thresholds<br>• Tailor risk process to match the risk challenge |

# THE TYPICAL RISK MANAGEMENT PLAN

## Purpose

As we consider the purpose and principles of risk management planning, it's clear that a number of important decisions need to be made during this step about how risk will be managed on our project. People need to know about what was decided, including our project team as well as other stakeholders, especially the project sponsor or business owner. This means we have to document the results of our decision-making and make it available to those who need to know.

Unsurprisingly, the main output of risk management planning is a Risk Management Plan (RMP)! This is one of the main scoping and controlling documents of the project, forming part of the overall Project Management Plan. The purpose of the RMP is *to record the outcome of key decisions on how risk will be managed on this project.*

Putting the principles of risk management planning into practice and recording the results in a project document doesn't sound very exciting, and perhaps this is why many project teams omit this step in favour of getting on with risk identification. A well-known project proverb says, "To fail to plan is to plan to fail." While this may be overly simplistic (although memorable), it contains an important truth. Without clearly specifying the parameters of risk management on our project, we can't tell if we're doing a good job. The RMP provides a standard against which we can measure our risk process (and against which we could be audited!), and we can then make necessary adjustments. This wouldn't be possible if this important document were missing.

This is why our risk process needs to include risk management planning as a distinct step before risk identification, to ensure that it is not forgotten or overlooked. Provision of templates for the RMP will also help project teams to produce this key document, as long as the templates don't include too many boilerplate elements that lazy project managers can simply adopt without thought. A good RMP template will prompt the project manager with the required headings for content, but it will also contain gaps in all the crucial places, forcing them to think about the right way to complete the RMP for this project.

A project-based organisation may also wish to provide several RMP templates for use by its projects, because this document itself is a scalable element of the risk process. For small or simple projects, a short-format RMP may suffice, perhaps no more than one or two pages. Bigger or more complex projects will require more detail in their RMP, requiring use of a different template. The RMP may be included as a section within an overall Project Management Plan, or it may be a stand-alone document, depending on project documentation standards.

## Contents

Whether the RMP is long or short, embedded or stand-alone, it needs to cover the same basic content. This is driven by how we answer the main questions outlined in the principles above. A typical RMP will have the following headings:

- *Introduction.* This describes the purpose of the RMP, with essential document reference information including author, issue date, revision status, approvals, etc.
- *Project objectives.* In this section we explicitly list all project objectives (scope, time, cost, quality, etc.), explain which ones will be in scope for the risk process on this project, and comment on their relative priority in case of conflict.
- *Objectives and scope of risk process.* Here we describe the purpose of risk management for this project, and clearly state the scope of the risk process, defining what is in scope and what is out. For example, we may wish to focus the risk process on a specific area of the project, maybe performing a technical risk assessment, or exploring aspects of contractual uncertainty, or examining which parts of the supply chain are most risky. This scope definition indicates which project objectives are considered to be "at risk", allowing us to identify and manage risks to these objectives in particular.

    As part of the scoping statement, we should list the types of risk which the risk process is expected to address for this project. This may be presented as a simple list of risk categories, or it may be structured in a hierarchical Risk Breakdown Structure (RBS)[*], either using a generic RBS or a project-specific version. An example RBS is shown in Figure 4-1.

---

[*] (Hillson, 2002a, b)

| RBS Level 0 | RBS Level 1 | RBS Level 2 | |
|---|---|---|---|
| 0. All Risks | 1. Technical Risk | 1.1 | Scope definition |
| | | 1.2 | Requirements definition |
| | | 1.3 | Technical processes |
| | | 1.4 | Technology |
| | | 1.5 | Technical interfaces |
| | | 1.6 | Design |
| | | 1.7 | Performance |
| | | 1.8 | Reliability and maintainability |
| | | 1.9 | Safety and security |
| | | 1.10 | Test and acceptance |
| | 2. Management Risk | 2.1 | Project management |
| | | 2.2 | Programme/portfolio management |
| | | 2.3 | Operations management |
| | | 2.4 | Organisation |
| | | 2.5 | Resourcing |
| | | 2.6 | Communication |
| | | 2.7 | Information |
| | | 2.8 | Health, Safety, and Environment |
| | | 2.9 | Quality |
| | | 2.10 | Reputation |
| | 3. Commercial Risk | 3.1 | Contractual terms and conditions |
| | | 3.2 | Warranties and liabilities |
| | | 3.3 | Internal procurement |
| | | 3.4 | Suppliers and vendors |
| | | 3.5 | Subcontracts |
| | | 3.6 | Client/customer stability |
| | | 3.7 | Partnerships and joint ventures |
| | | 3.8 | *Force majeure* |
| | | 3.9 | Dispute resolution/arbitration |
| | | 3.10 | Intellectual Property |
| | 4. External Risk | 4.1 | Legislation |
| | | 4.2 | Exchange rates |
| | | 4.3 | Site/facilities |
| | | 4.4 | Environmental/weather |
| | | 4.5 | Competition |
| | | 4.6 | Regulatory |
| | | 4.7 | Political/Country |
| | | 4.8 | Country |
| | | 4.9 | Social/demographic |
| | | 4.10 | Pressure groups |

**Figure 4-1**   Example Risk Breakdown Structure (RBS)

- *Risk process tailoring aspects.* This important section of the RMP outlines how the risk process will be implemented on this project. If the organisation has a standard risk process, then we should describe any deviations from the typical approach. Particular tailoring aspects might include:
    - Whether or not quantitative risk analysis will be used on this project, and if so, how and when
    - Which tools and techniques will be used for each step in the risk process
    - What level of risk reporting is required
    - How often risk reviews and updates will be undertaken

    The RMP section on risk reporting should include defining the level of detail required for the risk register on this project, where risk data is held in a consistent and usable format. Like other elements of the risk process, the risk register is not one-size-fits-all. We may choose a simple format risk register held in Microsoft® Excel® or a relational database, containing just the essential characteristics of each identified risk. Or we may need more detail about our risks, and decide to store this in a proprietary or custom risk database. Perhaps our organisation has a preferred risk tool that we should use, either on a project-by-project basis or containing risk data for multiple projects across the enterprise. Maybe we want to restrict access to a small number of project staff, or perhaps we need multi-user access across a number of locations. It should also cover the type of reports required, including purpose, content, format, frequency, distribution, etc. Table 4-2 details the type of risk data that can be held in a risk register, with an indication of the minimum data requirement, as well as what might be in a typical risk register and what a more detailed version might hold.

    The RMP will also define the level of risk reporting required for this project, and alternative risk report formats are discussed in Chapter 10.
- *Risk-related roles and responsibilities.* This is where we state who is responsible for the various elements of the risk process for this project, and describe their contribution, possibly using a Responsibility Assignment Matrix. Where possible, we should use the names of individuals rather than job titles, to encourage ownership. A summary of the risk-related responsibilities of key players is presented in Table 4-3.
- *Project-specific definitions of probability and impacts.* It is very important to define the terms to be used for qualitative assessment of risks on this particular project.* These definitions must reflect our understanding of stakeholder risk appetite, translating this into measurable indications of "how much risk is too much risk". It is common to use the terms "High, Medium, Low" to describe levels of probability and impact for a simple

---

* (Association for Project Management, 2008)

**Table 4-2 Risk Data Held in the Risk Register**

| Data Field | Minimum | Typical | Detailed |
|---|---|---|---|
| **Header Information** | | | |
| Project data (project title, project manager, client, project status) | | X | X |
| Document data (Risk Register issue number, date, approvals) | | X | X |
| Date of most recent risk review | | X | X |
| Date of next risk review | | | X |
| **Risk Identification Data** | | | |
| Unique risk identifier | X | X | X |
| Date identified | | X | X |
| Name of person identifying risk | | | X |
| Threat/opportunity indicator | | | X |
| Short risk title | X | X | X |
| Full risk description (cause/risk/effect) | X | X | X |
| Explanatory comments | | | X |
| Risk status (for example draft, active, closed, occurred etc.) | | X | X |
| **Qualitative Risk Assessment Data** | | | |
| Probability/frequency of occurrence (current, pre-response)<br>• Qualitative (for example High, Medium, Low)<br>• Quantitative (for example % range, frequency) | X | X<br>X | X<br>X |
| Impact on each project objective (current, pre-response)<br>• Qualitative (for example High, Medium, Low)<br>• Quantitative (for example, three-point estimates of days, dollars, technical performance measures) | X | X | X<br>X |
| Overall risk ranking<br>• Red/Yellow/Green (or similar)<br>• Risk Score (calculated from probability and impact)<br>• Risk Score (using more complex algorithm) | X | X<br>X | X<br>X<br>X |
| Other prioritisation parameters (urgency, manageability, detectability, proximity etc.) | | | X |
| Risk source (Risk Breakdown Structure) element) | | X | X |
| Project area affected (WBS element, CBS element, etc.) | | | X |
| Impact window | | | X |
| Action window | | | X |
| Related risks | | | X |

*(Continued on following page)*

Table 4-2 Risk Data Held in the Risk Register (*Continued*)

| Data Field | Minimum | Typical | Detailed |
|---|---|---|---|
| **Risk Response Data** | | | |
| Risk owner | X | X | X |
| Risk response strategy | X | X | X |
| Risk actions with Action owners | | X | X |
| Other data on risk responses and actions (for example schedule, budget, completion criteria) | | | X |
| Probability/frequency of occurrence (post-response)<br>• Qualitative (for example, High, Medium, Low)<br>• Quantitative (for example, % range, frequency) | | | X<br>X |
| Impact on each project objective (post-response)<br>• Qualitative (for example, High, Medium, Low)<br>• Quantitative (for example, three-point-estimates of days, dollars, technical performance measures) | | | X<br>X |
| Fallback plans | | | X |
| Risk response status (overall) | | X | X |
| Risk action status (detailed) | | | X |
| Secondary risks arising from risk responses | | | X |

project, or "Very High, High, Medium, Low, Very Low" for more complex projects. These subjective labels need to be turned into project-specific quantifiable definitions that the project team can use when prioritising risks, and the definitions are included in the RMP as a key part of the way risk will be assessed and managed on this project. The RMP should also define the prioritisation matrix to be used, usually called the Probability-Impact Matrix (P-I Matrix), defining how combinations of probability and impact ratings are translated into risk priority levels (either red/yellow/green, or Level 1/2/3, etc.). Examples of definitions of probability and impact levels and an example P-I Matrix are given in Table 4-4 and Figures 4-3 and 4-4.

- *Project-specific risk categorisation frameworks.* We may wish to group identified risks together to identify concentrations of risk in particular areas of the project. If this type of analysis is appropriate, the RMP will list the categorisation frameworks to be used. This will include the RBS, which lists potential causes of risk (Figure 4-1), but other project hierarchies may also be included—for example, the Work Breakdown Structure (WBS), Cost Breakdown Structure (CBS), Organisational Breakdown Structure (OBS), etc.

**Table 4-3 Risk-Related Roles and Responsibilities**

**Project Sponsor**—has overall accountability for the project and for delivering its benefits. The Project Sponsor must ensure that resources and funds are provided to the project for risk management. The role of the Project Sponsor in relation to risk management will include:
- Actively supporting and encouraging the implementation of risk management on the project
- Setting and monitoring risk thresholds and ensuring these are translated into acceptable levels of risk for the project
- Approving the Risk Management Plan prepared by the Project Manager
- Attendance at risk workshops, identification of risks and ownership of risks
- Reviewing risk outputs from the project with the Project Manager to ensure process consistency and effectiveness
- Reviewing risks escalated by the Project Manager which are outside the scope or control of the project or which require input or action from outside the project
- Taking decisions on project strategy in the light of current risk status, to maintain acceptable risk exposure
- Ensuring adequate resources are available to the project to respond appropriately to identified risk
- Releasing "management reserve" funds to the project where justified to deal with exceptional risks
- Regularly reporting the risk status of the project to senior management

**Project Manager**—has overall responsibility for delivering the project objectives in a way that allows the intended benefits to be achieved. The Project Manager must ensure that risks are identified and managed through effective risk management. The Project Manager reports to the Project Sponsor. The role of the Project Manager will include:
- Determining acceptable levels of risk for the project in consultation with the Project Sponsor
- Preparing the Risk Management Plan in conjunction with the Project Sponsor
- Ensuring that the risk process is implemented properly on the project
- Participating in risk workshops, review meetings, and identifying and owning risks
- Approving risk response plans and their associated risk actions prior to implementation
- Applying project contingency funds to deal with identified risks that occur during the project
- Overseeing risk management by subcontractors and suppliers
- Regularly reporting the risk status of the project to the Project Sponsor
- Highlighting to senior management any identified risks which are outside the scope or control of the project, or which require input or action from outside the project, or where release of "management reserve" funds might be appropriate

**Risk Facilitator**—(where present) has responsibility for overseeing and managing the risk process on a day-to-day basis. The Risk Facilitator reports to the Project Manager, although this role may be fulfilled by the Project Manager on smaller projects. The role of the Risk Facilitator will include:
- Preparation of the Risk Management Plan
- Facilitation of risk workshops and risk reviews at which risks will be identified and assessed
- Creation and maintenance of the Risk Register
- Liaising with Risk Owners to determine appropriate risk responses and reviewing progress on their implementation
- Ensuring the quality of risk data
- Analysing data and producing risk reports
- Coaching and mentoring team members and other stakeholders on aspects of risk management

*(Continued on following page)*

Setting the Scene for Opportunity Management 63

**Table 4-3 Risk-Related Roles and Responsibilities (*Continued*)**

| |
|---|
| **Risk Owner**—appointed by the Project Manager as the best person to manage an identified risk. The Risk Owner's role is temporary in that once a risk has been closed their role ceases. A Risk Owner can be a member of the project team, a stakeholder who is not part of the project team, or a specialist from outside the project. The role of the Risk Owner will include:<br>• Development of responses to risks in the form of risk actions which they will assign to Action Owners<br>• Monitoring the progress of Action Owners in implementing risk responses<br>• Reporting progress on risk responses via the Risk Register |
| **Project Team Members**—responsible to the Project Manager to follow the risk process. The role of Project Team Members will include:<br>• Participating actively in the risk process, proactively identifying and managing risks in their area of responsibility |

**Table 4-4 Example Risk Assessment Criteria for Likelihood and Impact**

| Scale Point | Likelihood | | Impact | | |
|---|---|---|---|---|---|
| | Probability | Frequency | Schedule | Cost | Performance |
| Very High | >50% | At least once each month | Threat: >6 months delay<br>Opportunity: >2 months early | Threat: >$250K cost growth<br>Opportunity: >$100K cost savings | Very significant impact on overall functionality (±) |
| High | 31–50% | Once every 1–3 months | Threat: 3–6 months delay<br>Opportunity: 1–2 months early | Threat: $101–250K cost growth<br>Opportunity: $50–100K cost savings | Significant impact on overall functionality (±) |
| Medium | 11–30% | Once every 3–12 months | Threat: 1–3 months delay<br>Opportunity: 2–4 weeks early | Threat: $51–100K cost growth<br>Opportunity: $10–50K cost savings | Some impact on key functional areas (±) |
| Low | 5–10% | Once every 12 months or more | Threat: 1–4 weeks delay<br>Opportunity: 1–2 weeks early | Threat: $10–50K cost growth<br>Opportunity: $5–10K cost savings | Minor impact on overall functionality (±) |
| Very Low | <5% | Once during the project lifetime | Threat: <1 week delay<br>Opportunity: <1 week early | Threat: <$10K cost growth<br>Opportunity: <$5K cost savings | Minor impact on secondary functions (±) |

# UNDERSTANDING RISK APPETITE AND RISK THRESHOLDS

As with other elements of the risk process, the principles of this risk management planning step apply equally to both threats and opportunities. Similarly, the content of the RMP remains the same for a threat-only risk process or a broader approach covering both threats and opportunities. There are, however, several aspects that require slight modification for opportunities, especially if an organisation is not used to tackling them through the risk process.

The first of these areas is *risk appetite*.* This can be defined as, "*the tendency of an individual or group to take risk in a given situation*" and is related to the riskiness of that situation, the level of potential rewards available, and the previous experience of the individual or group. Risk appetite is an internal drive, and so it is not easy to measure or monitor (similar to physical appetite or hunger). This means that risk appetite must be translated into something external that can be measured (in the same way that we express hunger in terms of an amount of food we'd like to eat). Risk appetite is expressed in measurable *risk thresholds*, which represent *upper and lower limits of acceptable uncertainty against each objective*.

It is relatively easy for project stakeholders to express their risk appetite in negative terms.† Taking the example of the objective to complete the project by a specified end-date, stakeholders can tell us whether they would be comfortable for a project to be delivered late by one day, one week, or one month. The level of delay that would be OK indicates the lower boundary of the risk threshold around the schedule objective. A project sponsor, business case owner, or client should be able to say what level of delay is completely unacceptable and would lead to project cancellation, and this gives us the upper boundary of the schedule risk threshold. This range of acceptable negative variation around the project schedule objective is easy to conceptualise and express, and the same is true for other project objectives.

But stakeholders often have more difficulty when considering positive variations. Surely any upside is welcome, the more the better? In fact, this is not usually true. For example, for project schedule, delivering the project a day early may be insignificant, whereas a week or a month might be seen as advantageous. But what if it were possible to deliver in half the time? This might be a saving too far, as staff have been allocated to this project and we might have trouble redeploying them, or we may have contractual delivery dates from suppliers or subcontractors, or perhaps we've agreed to meet set regulatory milestones. The same is true of the project budget objective, where a certain degree of underspend is welcome, but if the project were completed in half the budget, then resources would have been committed unnecessarily to this project that could have been used elsewhere. As a result, there is a level of positive saving that would be unhelpful, and this represents the upper level of acceptable variation.

In other words, when setting risk thresholds for a project, stakeholders need to consider both upside and downside variations on each objective. This will require careful guidance from a skilled facilitator who understands the nature of risk appetite and risk thresholds.

---

* (Hillson & Murray-Webster, 2012)
† (Hillson, 2016b)

In an ideal world, considering the risk appetite of project stakeholders will be done in the context of an overall understanding of corporate risk appetite and a set of defined risk thresholds around strategic corporate objectives. A risk-mature organisation will understand "how much risk is too much risk" for the entire enterprise, and it will be able to roll these strategic risk thresholds down throughout the organisation, to be used at lower levels when determining "how much risk is too much risk" at each level.

Sadly, not many organisations are this mature. Instead project teams are left to work out for themselves where risk thresholds should be set, in discussion with their project sponsor, business owner, and other key stakeholders. In this case, we need to start by having a careful, facilitated discussion with our stakeholders, either individually or as a group. The first step is to explore their risk appetite for the project as a whole, and for each specific project objective. Because risk appetite is an unmeasurable internal drive to take risk in pursuit of rewards, this discussion will necessarily be qualitative in nature. The facilitator asks each stakeholder to place themself on a scale of risk appetite, perhaps using Very Low/Low/Medium/High/Very High. Each level of risk appetite translates into a degree of acceptable variation, either for the project as a whole, or for the objective under consideration. Very Low risk appetite means that the stakeholders would be uncomfortable to see any significant variation, where Very High risk appetite corresponds to being prepared to accept wide variations on potential project outcomes. Figure 4-2 illustrates this graphically (not drawn to scale). The $y$-axis represents the degree of variation linked to each level of risk appetite. The upper part represents the possibility of positive variation (reduced duration or cost, improved quality, better performance etc.), and the lower part shows negative variation (late or overspent, lower quality or performance).

When we understand the risk appetite of project stakeholders, then we can work with the stakeholders to translate this into measurable risk thresholds around each project objective. These give an upper and lower boundary of acceptable variation for project duration or delivery date, outturn cost, performance parameters, etc. When risk thresholds are set for the overall project, as measurable upper and lower limits of acceptable uncertainty against each objective, these can be translated into specific risk criteria for use in qualitative assessment of both threats and opportunities.

## DEFINING RISK ASSESSMENT CRITERIA

It is common to use two key dimensions of risk when prioritising them, namely, the *probability* that the risk will occur, and the *impact* it would have on one or more objectives if it did occur. These two dimensions are directly linked to

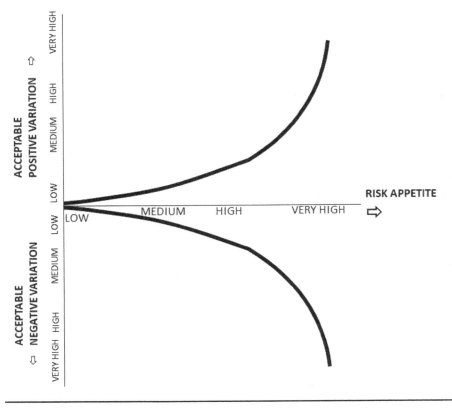

**Figure 4-2** Risk Appetite Levels and Degree of Variation

our simple definition of risk as "*uncertainty that matters*", because probability is a way to describe the uncertainty associated with each risk, and impact tells you how much it matters. There are other characteristics of a risk which can be used for prioritisation in addition to probability and impact[*], and these are discussed in Chapter 6. But all risks are uncertain, and all risks matter, so the main prioritisation approach is based on these two factors.

It is important to clarify at this point exactly what is meant by the term "probability" when we use it during risk assessment. In the risk process, probability is not an objective statement of the chance that a random event will occur. In reality, each risk either will or will not occur. The problem is that we don't know whether it will or it won't occur. The uncertainty does not lie in the

---

[*] (Association for Project Management, 2008)

randomness of the risk's occurrence or not; instead it is in our level of knowledge. This means that when we talk about "probability" as a measure of uncertainty, we are actually describing our level of confidence that we can predict the occurrence of a risk accurately. We're expressing how likely we are to be right when we say that we believe the risk will occur. If we assess probability as Low, it means that we are not confident in our prediction. Assigning a percentage range to this assessment, say 5–10%, means that we think there is a 90–95% chance that we're wrong. Similarly, if we say that probability is High, meaning >80%, we believe that we are almost certain that the risk will occur, and there is less than 20% chance that we turn out to be mistaken.

The difficulty is that we use the term "probability of occurrence" quite loosely in the field, allowing people to think they are required to estimate the actual chance that a risk will happen, which is quite hard. For example, we actually don't know whether a key supplier might go out of business during our project. But we can estimate how confident we feel when we say, "Yes, I think that will happen."

In order for the project team to assess risks objectively, they need an agreed framework that ranks risks in a consistent way. This requires definition of a common set of risk assessment criteria to be used in this project when assessing probability and impact. Each of these dimensions is described using a number of levels—at least three levels (Low/Medium/High), which is suitable for small or simple projects, or, more commonly, five levels (Very Low/Low/Medium/High/Very High) for larger or more complex projects. These qualitative terms have to be expressed more precisely if they are to be useful, so the project needs to have a set of scales which interprets each criterion level. In other words, we all need to know what is meant on this project when we talk about "medium probability" or "high cost impact".

These definitions are developed during the risk management planning step of the risk process, and they are recorded in the RMP. It is relatively easy to produce a *probability scale,* and often an organisation will provide a standard set of definitions to be used on all projects. For example, if we have decided to use a five-point scale, it might look like this:

- Very High probability = >50% confident that the risk will happen
- High probability = 31–50% confident that the risk will happen
- Medium probability = 11–30% confident that the risk will happen
- Low probability = 5–10% confident that the risk will happen
- Very Low probability = <5% confident that the risk will happen

Probability should only be used for one-off risks for which there is no data on previous occurrences, and all we can do is state our level of confidence that this risk might occur on our project. Risks which occur more than once are treated differently, by considering the frequency with which they occur. For example:

- Very High frequency = expected to occur at least once each month
- High frequency = expected to occur once every 1–3 months
- Medium frequency = expected to occur once every 3–12 months
- Low frequency = expected to occur once every 12 months or more
- Very Low frequency = expected to occur no more than once during the project lifetime

Though it is best practice to define scales for both probability and frequency to describe the uncertainty dimension (and perhaps to use the term "likelihood" as a generic label covering both), it is more common only to define a probability scale. This introduces problems when trying to assess the uncertainty associated with repeated uncertain events, such as the possibility that test hardware might break down during the trial period.

Once a set of probability scales (and possibly also frequency scales) have been developed to allow the uncertainty dimension to be assessed, it is time to turn to *impact scales,* which describe how much a risk matters if it occurs. Risks matter because if they occur, they would affect achievement of one or more project objectives. As a result, impact scales are required for each project objective, so that we can assess how much a risk might affect the project timeline and/or its budget and/or performance criteria, etc. These scales will allow us to translate the terms Very Low, Low, Medium, High, and Very High into numerical ranges, which can then form the basis of a consistent assessment of the impact dimension. Impact scales for threats define increasing levels of unwelcome or adverse impact, ranging from the insignificant (Very Low) to the intolerable (Very High). Opportunity impact scales define increasing levels of positive impact, ranging from insignificant (Very Low) to unmissable (Very High).

Sometimes project teams develop these impact scales intuitively, without any outside reference to tell them "how much risk is too much risk" on this project. But this is precisely the purpose of *risk thresholds*. Risk thresholds provide measurable upper and lower bounds of acceptable variation for each project objective, and these can be translated directly into risk assessment criteria for the impact dimension. The upper risk threshold of maximum acceptable positive variation defines the level of positive impact that would be regarded as the best possible. By contrast, the lower risk threshold of maximum acceptable negative variation defines the level of negative impact that would be regarded as totally unacceptable. These two limits should determine the Very High level of impact to be used when assessing opportunities and threats, respectively.

A simple process can be used to turn risk thresholds into risk assessment criteria for the impact dimension, as follows (assuming we are using a five-point scale):

- *Use the outer limits of the risk thresholds to define the highest level of impact in the risk assessment criteria* (Very High). For example, if the risk thresholds for project delivery date state that any delay more than six months could not be allowed, and that we could not imagine delivering more than two months early because of resource limitations, then we can define Very High schedule impact for threats as "more than six months delay", and Very High schedule impact for opportunities would be "delivering more than two months early".
- *Ask key project stakeholders to state the level of impact that would not cause any problems,* either because it is within the total float or budget reserve of the project, or because it would be insignificant in the overall context of the project. These insignificant impacts are used to define the lowest level of impact (Very Low). If a cost overrun of $10K would not be noticed, or nobody cares about a saving of only $5K, then a Very Low level of cost impact for threats would be set at "<$10K cost growth", and Very Low cost impact for opportunities would be "<$5K cost savings".
- Once the outer levels of the risk assessment criteria are set for impacts on each project objective, *spread the intervening levels (Low/Medium/High) between top and bottom values* to provide a fully defined impact scale.

This process is illustrated in Figure 4-3, starting at the top by defining Very High levels of impact for both threats (intolerable) and opportunities (unmissable), then moving to the bottom to set the Very Low levels (insignificant), then passing back up the scale to set increasing levels of impact (minor, significant, major).

Table 4-4 shows example definitions of likelihood (covering both probability and frequency) and impact for both threats and opportunities, using a five-point scale.

Although it is better to use the generic term "likelihood" for the uncertainty dimension of risks, and to include both probability (for one-off risks) and frequency (for repeated events) within that term, it is common practice to talk only about probability. While this is undoubtedly simpler, because it avoids the need to think too deeply about different types of uncertainty, it can cause project team members to stumble when trying to assess risks that are not one-off events. Nevertheless, we will follow common practice here, using the terms "probability and impact" as short-hand labels for the two key dimensions of risk (uncertainty and effect on objectives).

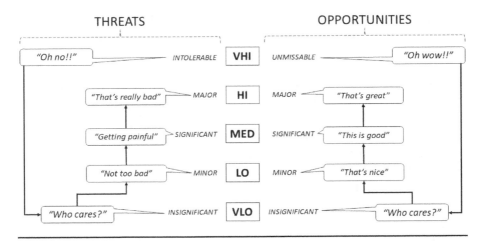

**Figure 4-3**  Defining Levels of Impact

## DEFINING RISK ASSESSMENT FRAMEWORK

When we have defined the probability and impact scales to be used for assessing risks, we need to explain how these two dimensions will be combined to allow the importance of each risk to be compared. This is where the two-dimensional Probability-Impact Matrix (P-I Matrix) is used, divided into prioritisation zones that allow each risk to be plotted into an appropriate cell that represents its overall ranking. It is common to have three prioritisation zones; often these are labelled Red/Yellow/Green (following the traffic-light model), or Level 1/2/3. Top-priority risks lie in the Red zone or are at Level 1, medium-priority risks are in the Yellow zone or Level 2, and low-priority risks are Green or Level 3. Where a more detailed risk assessment is required, the P-I Matrix can be divided into more than three prioritisation zones, using different colours (Black/Red/Orange/Yellow/Green) or levels (Level 1/2/3/4/5).

Each prioritisation zone includes a set of combinations of probability and impact that are defined as being of equal importance (all Red risks are equally red). The boundaries between these zones are defined in the RMP, to provide a consistent prioritisation framework for the project. The positioning of the boundaries is a direct indication of the risk thresholds for the project, and it is not arbitrary. We need to think about which combinations of probability and impact would constitute a top-priority risk, and which combinations would lead us to treat a risk as low-priority.

A typical P-I Matrix for prioritising threats is shown in Figure 4-4, which uses five-point scales for both probability and impact, with increasing probability on

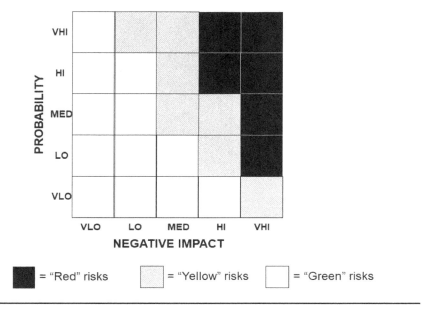

**Figure 4-4** Probability-Impact Matrix for Threats

the *y*-axis and increasing negative impact on the *x*-axis. (A smaller project may use only three or four points on the probability and impact scales, in which case the P-I Matrix is correspondingly smaller.) It is notable that the matrix is not symmetrical; in other words, the boundaries between zones do not run across the matrix at 45°. Instead we can see that the boundaries are skewed towards impact. This reflects the fact that we naturally wish to regard impact as more important than probability. To illustrate this preference, consider two threats:

- Risk A: probability = Very Low, impact = Very High
- Risk B: probability = Very High, impact = Very Low

It is natural to regard Risk A (a small chance of a disaster) as more important than Risk B (an almost inevitable minor inconvenience). This weighting of impact over probability is reflected in the asymmetrical zone boundaries on the P-I Matrix in Figure 4-4, where Risk A is shown as medium-priority (Yellow zone) and Risk B is low-priority (Green zone).

An identical matrix can be used to prioritise opportunities by simply using the *x*-axis to show increasing positive impact. It is, however, common to use a modified "mirror matrix" containing both threats and opportunities, but with the opportunity side flipped over (see Figure 4-5). This brings the two Red zones into the centre of the combined P-I Matrix, providing an area of

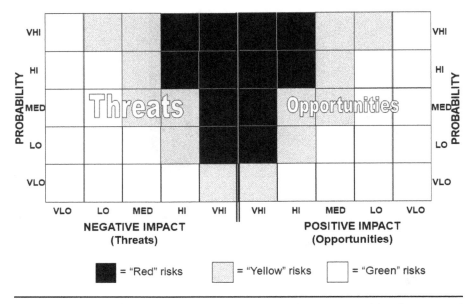

**Figure 4-5** Combined "Mirror" Probability-Impact Matrix for Threats and Opportunities

focus containing the worst threats and the best opportunities. These are the top-priority risks that the project needs to address. On the outer edges of the combined matrix are the two Green zones, containing low-priority threats and opportunities that are either unlikely to happen or that have small impacts on the project, or both.

The use of Red/Yellow/Green zones to prioritise opportunities is questioned by some, who ask "Why is a really good opportunity shown as Red? Surely Red means danger or bad?" However, this colour scheme is used for both threats and opportunities because it is based on the colours of traffic lights:

- *Red means Stop.* A threat in the Red zone is so bad that we cannot ignore it and keep going. We have to stop and consider this threat actively and find ways to neutralise it or reduce it. In the same way, an opportunity in the Red zone is so good that we cannot ignore it and keep going. We have to stop and consider this opportunity actively and find ways to capture it or make it more likely.
- *Yellow means Take Care.* Both threats and opportunities in this zone should be monitored regularly to ensure that their priority level doesn't change. We should be prepared to stop if necessary (if the traffic light changes to Red), or to continue on (if it becomes Green).

- *Green means Go.* Low-priority threats and opportunities do not need active attention or action, although they should be checked from time to time to ensure that they remain low priority. We can just carry on past them without changing direction.

If a project team are genuinely bothered by the use of Red/Yellow/Green for opportunities, they can easily replace them with a different set of colours, perhaps Gold/Silver/Bronze, or shades of blue.

## SUMMARY AND REFLECTION QUESTIONS

The novice project team member or project manager who isn't familiar with the principles of risk management might be forgiven for thinking that they could omit the first step in the risk process and get on with identifying risks. Why waste time planning the risk management approach for this project?

Even if we recognise the need to start with a planning step to define the scope and parameters of risk management, there's a temptation to quickly complete an RMP template to produce a document that ticks the box, allowing us to move on to the more interesting and important parts of the risk process.

This chapter has shown that we really need to spend time and effort on the risk management planning step. Decisions are made during this phase that determine which objectives are the focus of risk management, how we know what level of risk exposure would be tolerable and what is unacceptable, how we will identify risks, how we prioritise risks for further action, what the reporting requirements are, how often we update our risk assessment, and who is responsible for the various elements of the risk approach. Understanding the factors that underlie each of these decisions is not a trivial task, and risk management planning is essential if our risk approach is to be effectively tailored to the risk challenge of this particular project. Documenting our decisions in the RMP is merely the last part of this stage, in which we communicate the important parameters of the risk process for this project to our project stakeholders.

### Reflection questions

- Why is it important for the risk process to start with risk management planning?
- Why should we tailor the risk management approach for each particular project? Which elements of the risk process can be tailored, and how?

- What differences will there be in the RMP if we include both threats and opportunities in our risk approach, instead of just threats? And what elements of the RMP will remain the same?
- What's the difference between risk appetite and risk thresholds? How are risk thresholds related to risk assessment criteria? Why does this matter?

## THE NEXT STEP ("NOW WE'VE DEFINED THE PROCESS, WHERE ARE THE OPPORTUNITIES?")

Risk management planning is important because it sets the scene for the subsequent elements of the risk process, starting with risk identification. The RMP defines the parameters that are required if we are to be able to undertake the risk process successfully, including:

- The project objectives that are "at risk", allowing us to identify specific threats and opportunities that would affect those objectives
- The types of risk that we should be looking for, defined in the RBS
- The techniques that we can use to identify risks
- The assessment criteria that will form the basis for prioritising threats and opportunities

With these elements clearly defined and communicated to the project team and other stakeholders, we're now in a position to start looking for threats and opportunities. Cue the next step in the risk process—risk identification, and in particular, how do we find opportunities?

# Chapter 5
# Finding Opportunities

In the initiation stage of the risk management process, we set the scene for how risk will be managed on this project, and we record our tailoring decisions in the project's Risk Management Plan. Everyone now knows that we expect to identify and manage opportunities in this project, and we're ready to get started. The next step is to **find those opportunities**, alongside the threats. We're used to looking for downside risks, but how will we identify opportunities? This chapter outlines the purpose and principles of risk identification and briefly describes the main techniques used to identify threats. Some of these can be used or adapted to look for opportunities, but there are also specific techniques that intentionally set out to identify them.

If we're serious about managing opportunities, first we have to find them!

## PURPOSE AND PRINCIPLES OF RISK IDENTIFICATION

Many people believe that risk identification is the most important stage of the risk management process. After all, if we don't identify a risk, we can't manage it. Some (many?) unidentified threats will turn into problems that could have been avoided or at least minimised. The potential benefits and savings available from some (most?) unidentified opportunities will be missed.

It's easy to say that *the purpose of risk identification is to identify risks.* However, experience shows that this isn't as easy as it sounds. Often the risks identified in this stage are "the usual suspects", things that everyone knows about and expects, and that are dealt with through our standard project processes. We might call these "business-as-usual risks". Typical examples include:

- The customer might change the requirement.
- Key resources may not be available when required.
- Our deliverable may not match expectations of key stakeholders.

Each of these risks should be dealt with through our routine management of the project, using standard techniques such as change control, resource planning, and stakeholder engagement. Risk management should instead be focused on those risks that would not be covered by our standard operating procedures, and for which we have no existing process. If the response to a risk is that "Someone should do their job and follow the proper procedure", then it is not a risk—unless we suspect that someone might not do their job or that the proper procedure might be inadequate, which would indeed be a risk that needs managing.

In addition, we need to recognise that not all risks are currently knowable. It's not possible to see all risks from the perspective of our current position in the present. Some risks can only be identified later on in the project. There are several reasons for this:

- *Time dependence.* As we make progress on the project, we can see farther over the horizon, or around the next corner, where previously invisible risks might be lurking.
- *Action dependence.* Some risks arise from our decisions and actions, and these can't be identified until we make the decision or take the action.
- *Risk-response dependence.* In particular, our responses to identified risks might give rise to new risks which wouldn't exist if we didn't implement our chosen response. These are known as "secondary risks", not because they are less important, but because their existence depends on implementing the response to another risk.
- *Stakeholder dependence.* Risks can arise from the decisions or actions of "invisible stakeholders", who have the ability to affect the project but who are currently not recognised by those trying to identify risks.

In addition, there are two categories of risk that are very difficult to identify:

- *Random risks.* These are uncertain events or conditions that are inherently unknowable or unforeseeable, because they are the product of random chance.
- *Perceptually-concealed risks.* People are unable to perceive some risks due to their cognitive biases or preconceived paradigms—these risks are literally "inconceivable" to them.

The fact that we can only identify those risks that are currently knowable means that we need to return to the risk identification step regularly in order to find risks that were previously unknowable but that have emerged since the last time we looked. We should also be regularly challenging our preconceptions and sources of bias in order to broaden our view of potential risks.

It is therefore more accurate (and helpful) to say that *the purpose of risk identification is to identify knowable risks that otherwise would not be managed*. We might call these the "real risks", as opposed to generic "business-as-usual risks" that should be addressed by standard operating procedures, or dependent risks that might emerge in the future.

It is clearly vital for us to identify the real risks to our projects, so that we can prioritise and manage them proactively. As we approach the risk identification stage, there are several key *principles* to apply.

## All risks are uncertain

We often forget this self-evident truth when we try to identify risks. As a result, many of our so-called "risks" are actually certainties, including facts, constraints, requirements, issues, problems, etc. If we identify something that is not uncertain, then we can be sure that it is not a risk.

Of course, some certainties can give rise to risk, and we should examine them as potential causes of risk. But some certainties are simply things we have to live with (for example, a recruitment freeze in the company), part of the context of the project (strong competition), non-negotiable constraints and requirements (testing must be done at the client site), or issues and problems to be dealt with (three key staff have just resigned).

## Each risk must be linked to at least one objective

All risks are indeed uncertain, but not all uncertainties are risks. Many uncertainties don't matter to our project. Risk is "uncertainty that matters", and it matters because if it occurs it will affect the achievement of one or more objectives. If it's not possible to make an explicit link between a so-called risk and an objective, then it isn't a risk, it's just an irrelevant uncertainty that doesn't matter.

Ensuring that each identified risk is connected to at least one objective provides an important sanity check for our risk identification. As we seek to identify risks, we should constantly be asking ourselves "So what? Why does this matter? What difference would it make if it happened?"

The connection with objectives is also important later on in the risk management process, particularly in risk response planning and risk reporting. We'll see that the best person to own a risk and ensure that it is managed effectively is usually the person who owns the objective that would be affected if the risk occurred. Linking risk ownership to objectives provides an incentive to manage the risk effectively, because the risk matters to the risk owner. We can also use the connection between risks and objectives to structure risk reports, listing all the risks that might affect a given objective, using that information to determine the risk exposure of each objective, and targeting risk information to the people who most need to know.

## Consider different time perspectives

As we seek to identify risks, we can consider three distinct time perspectives: past, present, and future. We might characterise these as *looking back*, *looking around*, and *looking forward*. Each of these dimensions of time can provide useful information about the risks we might face in our project.

- *Past*. We can review historical data on previous completed similar projects (or earlier phases of our current project), looking at audits, post-project review reports, lessons-learned databases, risk checklists, case studies, or industry reports. By seeing what risks were identified in those cases, we might discover similarities with our own project. We might also find details of risk responses that worked particularly well before, which we could adopt or adapt to respond to our own risks.
- *Present*. The characteristics of our project should be examined carefully to reveal areas of uncertainty built into the scope, requirements, contractual terms, technical assumptions, business case, etc. Many risks that are specific to our own project will be hidden here, within the documentation that defines the project.
- *Future*. There are many creativity techniques that can be used to visualise potential futures that might come to pass for our project. These require us to suspend our current constraints and perceptions, and to think outside the box. A skilled facilitator can be particularly helpful here. As we peer ahead, using risk identification as a forward-looking radar to scan the future, we can spot areas of uncertainty that are heading in our direction.

Looking in these three different time dimensions to identify risks will require three different skillsets: *analysis* (of the past), *curiosity* (about the present), and *imagination* (for the future). These are embodied in various risk identification techniques, as we shall see.

## Use more than one risk identification technique

No single risk identification technique will be able to identify all the real risks that are currently knowable. This is partly due to the need to consider the three dimensions of time mentioned above—most techniques consider only a single time dimension. Some risk identification techniques are best performed by individuals, and others require the participation of a group. Some techniques target particular types of risk, such as internally generated risks or external risks or technical risks.

As a result, we should always use more than one risk identification technique, to ensure that we address all time perspectives, harness both individual and group perceptions, and cover all types of risk.

## Include multiple perspectives

In addition to taking input from both individuals and groups when identifying risks, it's important to include all key stakeholder perspectives. As a minimum, we need to hear from the project team, the project sponsor, and the customer. Depending on the type and size of our project, we may also need to include suppliers, subcontractors, consultants, technical specialists, procurement specialists, legal advisers, user representatives, regulators, lobby groups, etc. It may not be necessary to have representatives from each of these different stakeholder groups together in a single risk identification session, as long as their opinions and inputs are faithfully represented by someone.

If we only take one perspective on the risks faced by our project (usually the view of the project team), then we're likely to miss important risks that others would see clearly.

## Consider all potential sources of risk

Many people have no trouble identifying risks in areas where they have some experience or expertise. This is unsurprising, but not always helpful. Technical specialists will spot technical risks, procurement experts know how to find contractual risks, and project planners are used to identifying risks to schedule or budget. But the real risks to our project are likely to sit in those areas which lie outside our comfort zone, where we rarely look. We need to learn how to look in our blind spots, examining those areas which are less familiar to us.

Prompt lists can be useful in reminding us of risk categories that we might otherwise overlook. A more effective tool is to use a Risk Breakdown Structure (RBS), which provides a hierarchical framework of risk sources. A typical

RBS has four Level 1 risk sources: Technical, Management, Commercial, and External. These four main areas represent generic sources of risk that affect most types of project. The RBS is then taken to a lower level of detail, where Level 2 represents specific risk sources that are relevant to our organisation or to this type of project. In rare cases it might be useful to take the RBS one step further, to define Level 3 detailed risk sources, but two levels are usually sufficient for most projects. An example generic RBS is shown in the previous chapter (Figure 4-1).

We can use a tailored RBS to help us identify risks in a structured way. We might take RBS Level 2 headings as an agenda for our risk identification workshop, or as a prompt list, or to shape the questions in a risk interview. Or we might use the RBS as a sanity check after using another risk identification technique, to make sure that we haven't missed anything. The RBS reminds us about potential blind spots where risks might be lurking outside our comfort zone, and it encourages us to go looking.

## Repeat risk identification throughout the project

The fact that not all risks are immediately visible at any given point in time means that we need to stay vigilant and keep looking for risk at regular intervals throughout the project. The meaning of "regular intervals" will be different for each project, but they should align with the frequency of project progress reviews. For projects with a rapid pace of change or development, risk identification sessions may be needed often, perhaps weekly. Longer duration projects with a slower lifecycle may only need to review their risk exposure every three to six months, or even annually.

In addition to regular reviews, identification of newly visible risks should occur at key points in the project lifecycle, including phase transitions, as well as when any major change is proposed (for example, a change in scope, introduction of a new development approach, or significant re-planning).

**Table 5-1 Purpose and Principles of Risk Identification**

| Purpose | To identify knowable risks that otherwise would not be managed |
|---|---|
| Principles | • All risks are uncertain<br>• Each risk must be linked to at least one objective<br>• Consider different time perspectives<br>• Use more than one risk identification technique<br>• Include multiple perspectives<br>• Consider all potential sources of risk<br>• Repeat risk identification throughout the project |

## Summary

The purpose and principles of risk identification are summarised in Table 5-1.

## SEPARATING RISK FROM NON-RISK USING RISK METALANGUAGE

Another challenge in identifying "real risks" is to avoid confusion between causes of risk, genuine risks, and the effects of risks.* Many risk registers include so-called "risks" that are not uncertain, but that might give rise to risks. Although it's important and helpful to know about these causes, they are not risks. In the same way, risk registers sometimes contain descriptions of risk impacts and call these "risks".

How do these three differ?

- *Causes* are definite events or sets of circumstances which exist in the project or its environment, and which give rise to uncertainty. Examples include the requirement to implement the project in a developing country, the need to use an unproven new technology, the lack of skilled personnel, or the fact that the organisation has never done a similar project before. Causes themselves are not uncertain, because they are facts or requirements, so they are not the main focus of the risk management process.
- *Risks* are uncertainties which, if they occur, would affect the project objectives either negatively (threats) or positively (opportunities). Examples include the possibility that planned productivity targets might not be met, that interest or exchange rates might fluctuate, that client expectations may be misunderstood, or that a contractor might deliver earlier than planned. These uncertainties should be managed proactively through the risk management process.
- *Effects* are unplanned variations from project objectives, either positive or negative, which would arise as a result of risks occurring. Examples include being early for a milestone, exceeding the authorised budget, or failing to meet contractually agreed performance targets. Effects are contingent events, unplanned potential future variations which will not occur unless risks happen. As effects do not yet exist, and indeed they may never exist, they cannot be managed directly through the risk management process.

The relationship between cause, risk, and effect is illustrated in Figure 5-1. This reflects the simplest case, in which one cause leads to a single risk which

---

* (Hillson, 2000)

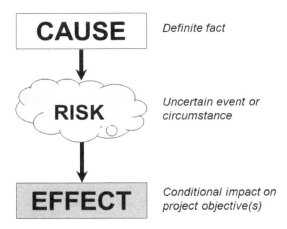

**Figure 5-1** Cause-Risk-Effect Relationships

in turn could have just one effect, though of course reality can be considerably more complex.

Including causes or effects in the list of identified risks obscures genuine risks, which may not receive the appropriate degree of attention they deserve. So how can we clearly separate risks from their causes and effects? One way is to use a structured description to provide a three-part risk statement, or *risk metalanguage,* as follows:

"As a result of <one or more definite causes>,
    <uncertain event> may occur,
        which would lead to <one or more effects on objective(s)>."

Examples of risks described using risk metalanguage are shown in Table 5-2.

The use of risk metalanguage ensures that risk identification actually identifies risks, distinct from causes or effects. Without this discipline, risk identification can produce a mixed list containing risks and non-risks, leading to confusion and distraction later in the risk process.

## TYPICAL TECHNIQUES FOR IDENTIFYING THREATS

There are many techniques commonly used to identify threats in projects. These can be divided into three groups, based on their time perspective (focused on past/present/future), as discussed above.[*] Details of how to use these techniques

---

[*] (Project Management Institute, 2009)

Table 5-2 Example Risk Descriptions Using Risk Metalanguage

| Cause (A Definite Fact) | Risk (An Uncertain Event or Set of Circumstances) | Effect (A Direct Impact on a Project Objective) |
|---|---|---|
| As a result of using novel hardware . . . | . . . unexpected system integration errors may occur . . . | . . . which would lead to overspend on the project. |
| Because our organisation has never done a project like this before . . . | . . . we might misunderstand the customer's requirement . . . | . . . which would mean that our solution would not meet the quality acceptance criteria. |
| We have to outsource production . . . | . . . [so] we may be able to learn new practices from our selected partner . . . | . . . which would lead to increased productivity and profitability. |
| Because we have no experience of using this technology . . . | . . . we might not have the necessary skilled staff to carry out the design work . . . | . . . which would lead to a delay in the project while we train our staff or recruit new skilled staff. |
| The project is planned to take place during the summer . . . | . . . [so] we may be able to recruit additional skilled student labour . . . | . . . which would save time on all activities that take place over that period. |
| Because there are three other projects taking place in the same time frame . . . | . . . we may be able to utilise skilled staff as they become available from another project . . . | . . . which would allow us to deliver early to the customer. |

are widely available in textbooks, training courses, and online resources, and they are not described here because our focus is on identifying opportunities.

## Past-focused techniques

These techniques are based on reviewing previous knowledge or experience, and then comparing what happened before with the current project to find points of similarity or relevance.

- Checklist
- Industry knowledge base
- Post-project reviews
- Lessons-learned database
- Historical information

## Present-focused techniques

Here the focus is on the current project, examining its characteristics and parameters to expose inherent uncertainties, inconsistencies, contradictions, or conflicts.

- Assumptions analysis
- Cause-and-effect diagram (also called Ishikawa or fishbone diagrams)
- Document review
- FMEA/Fault tree analysis
- Influence diagram
- System dynamics modelling

## Future-focused techniques

These techniques harness the creativity of the project team and other stakeholders to imagine what might happen in the future of the project.

- Brainstorming
- Delphi (expert review) technique
- Interviews
- Futures thinking
- Nominal Group Technique
- Prompt list
- Questionnaire
- Scenario analysis
- Visualisation

# MODIFYING THREAT TECHNIQUES TO FIND OPPORTUNITIES

Although it is common for the techniques listed above to be focused in practice on finding threats, some of them can also be used to identify opportunities, with some modification.

## Modifying past-focused techniques

Among the techniques listed in the previous section, it is often difficult to modify past-focused techniques so that they find opportunities. This is particularly true in the early stages of an organisation beginning to include opportunities in the risk process. In this case, they have no previous experience of identifying opportunities on which they can draw. Checklists will contain only threats, and post-project reviews or lessons-learned databases will also not include anything about opportunities. As time passes, of course, experience in identifying opportunities can be captured in these historical resources, updating checklists and lessons-learned databases to include upside risks. Until then, these techniques are unlikely to be particularly useful to support opportunity identification.

## Modifying present-focused techniques

Some present-focused techniques might be easier to adapt to find opportunities.

- An *influence diagram* should include both negative and positive drivers and influences, allowing upside variations in its constituent entities and outputs. The same is true of a *system dynamics model*, which is built on the same principles as an influence diagram, but with a higher level of detail and complexity.
- *Document review* can easily be used to look for uncertainties with potential positive impacts (opportunities) as well as downside risks (threats), by scanning key project documentation such as the contract, technical specifications, scope statement, business case, project charter, work instructions, etc. However, this is likely to be difficult, because we are often conditioned by habit to look for risks in a particular way, and we might easily miss opportunities during a structured document review.
- One commonly used risk identification technique that can be adapted to find opportunities is *assumptions analysis*. We can find threats to the project by examining various types of assumptions (scope, technical, estimating, planning, etc.) and asking two key questions:
  1. *How likely is this assumption to be false?*
     o Some assumptions are safe and are extremely unlikely to prove false—for example, we assume that our company will remain in business during the lifetime of this project. Others are less stable and could well be mistaken—such as, we assume that we have a full understanding of the customer requirement.
  2. *If it was false, how much would it affect the project's ability to achieve its objectives?*
     o Some assumptions would not matter if they turned out to be wrong—for instance, the client will prefer project reports to be delivered in written form. Others would matter a great deal—for example, staff motivation will remain at its current high level.
  
  Where an assumption is found to have a realistic chance of being false and it also could have a significant impact on the project, then that assumption is clearly risky. Because most assumptions are optimistic, a false assumption poses a threat to the project.

This technique can be extended to find opportunities by addressing project constraints in the same way as we use assumptions analysis to find threats. *Constraints analysis* involves testing various types of constraint (technical, contractual, resourcing, scheduling, etc.) and asking two questions that mirror those used in assumptions analysis:

3. *How likely is this constraint to be false?*
    o Some constraints are fixed and are very unlikely to change—for instance, the budget is fixed at USD $5M. But there may be some flexibility in other stated constraints—as in, all deliverables must be signed off by two senior managers. And others may just be plain wrong—such as, the project must end on 31 December this year.
4. *If it was false, how much would it affect the project's ability to achieve its objectives?*
    o It may be that some constraints could indeed be relaxed or removed, but that this would not have any significant effect on the project—for example, the client said that their staff would only be available during working hours, and even if we could contact them in the evenings, it wouldn't be useful. But we might find some flexible constraints whose removal would be really helpful to the project—as in, no contact is allowed between the project team and user representatives. But if we could . . .

A false constraint is one which is imposed on the project, either externally or internally, but which might possibly be relaxed or removed. If changing the constraint would affect the project, then the constraint is risky. Because most constraints have an inhibiting effect on the project, a changeable constraint might offer an opportunity to the project.

A simple table format can be used to test project assumptions and constraints in a combined approach, as shown in Table 5-3. This is used as follows:

o List all project assumptions and constraints in the left-hand column. (Hint: These can often be found in standard project documentation, such as the project charter, business case, contract, statement of work, technical requirements specification, estimating database, etc.)

o Identify whether each assumption or constraint might prove false (Yes/No), and whether it might affect the project if actually did turn out to be wrong (Yes/No).

o Where both answers are Yes, mark the assumption/constraint as a risk.

o Rewrite risky assumptions/constraints using risk metalanguage.

Several worked examples of this process are shown in Table 5-3.

Table 5-3 Example Assumptions and Constraints Analysis Worksheet

| ID | Assumption/ Constraint | Could Be False (Y/N) | Would Matter If False (Y/N) | Convert to Risk (Y/N) | Risk Description (Cause/Risk/Effect) |
|---|---|---|---|---|---|
| A1 | All key staff will be available when required. | Yes | Yes | Yes | The project plan assumes full resource availability. Some key staff may not be available when required, leading to delays in project execution. |
| A2 | Project documentation will be delivered in English. | No | n/a | No | n/a |
| A3 | Our main supplier will stay in business during the project lifetime. | Yes | Yes | Yes | Because we are dependent on one main supplier, if that supplier goes out of business during the project, we would be delayed while we found an alternative supplier. |
| C1 | User training will be delivered every Monday for four weeks after delivery. | Yes | No | No | n/a |
| C2 | Detailed design work cannot start until all high-level design is approved by client. | Yes | Yes | Yes | The client requires prior approval of high-level design before detailed design can proceed. The client may agree to approve and release low-risk design elements for development before full design approval. This would save time in the development phase. |
| C3 | The development team cannot be increased due to a recruitment freeze. | Yes | Yes | Yes | The current recruitment freeze prevents additional staff being hired. If an urgent need for a specific skill became evident, it might be possible to negotiate a one-off recruitment, preventing project delay. |

- Another present-focused technique that can be adapted for opportunities is *fault tree analysis*. This starts by stating a possible failure state of the project (delivering late, failing to meet requirements, being over budget, etc.), and determining what factors might lead to that happening. We then look for uncertainties that could strengthen the drivers of failure, and these are threats to be managed in order to minimise the chances of failure occurring.

A parallel technique can be used to identify project opportunities, by building a benefit tree instead of a fault tree. *Benefit tree analysis* starts from a success state (saving time, delighting the customer, coming in under budget, etc.) and asks what drivers might lead to this positive outcome. Uncertainties that could strengthen the drivers of success are opportunities to be managed in order to maximise the chances of success occurring. An example benefit tree is illustrated in Figure 5-2.

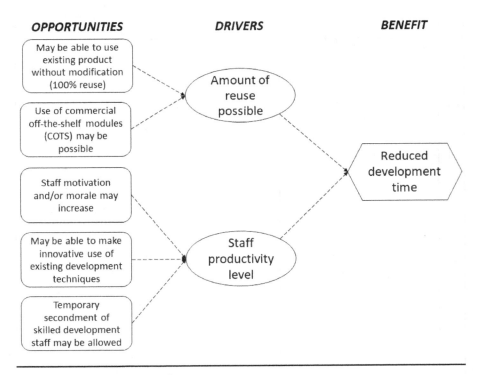

Figure 5-2   Example Benefit Tree

## Modifying future-focused techniques

Finally in this section, we consider how future-focused risk identification techniques might be modified to find opportunities. Here the main problem is one of mindset. Each of these forward-looking techniques requires people to exercise a degree of creativity and imagination, thinking about things that have not yet happened, but which realistically might happen, and which would then lead to an effect on project performance and outcomes. But we can only imagine things that lie within our existing frame of reference or worldview. If we are convinced that all risks are bad (threats), and that the only good risk is a dead risk, then our creative energies will not be harnessed towards imagining potential good things that might help us achieve project objectives (opportunities).

Whether we are brainstorming, being interviewed, completing a questionnaire, or visualising future scenarios, the absence of opportunities from our worldview means that they are literally inconceivable for us. Future-focused risk identification techniques will only become useful for identifying opportunities when we know that such things exist, that they are important, and that we can identify and manage them proactively.

It is possible to overcome this threat-focused mindset using effective facilitation, and some specific techniques can help to unlock people's creativity to think more positively. These are particularly useful when *brainstorming*, but they also work well in *interview* settings. The use of a structured workshop or interview to identify threats is well established, and people think naturally about threats in these sessions. Asking, "What could go wrong?" immediately results in a flood of ideas and suggestions. But people need more guidance to help them think about possible upside risks, uncertainties that would have a positive effect on achievement of objectives if they happened. Several creative approaches have been proposed:

- *Three wishes.* People familiar with children's stories will know about the fairy godmother or the genie in a bottle who offers three wishes to the lucky hero or heroine. We can run a "three wishes" workshop or interview in which participants are invited to wish for things that would make their life easier, or help them do their job better, or lead to a successful project outcome. Some of these wishes can be developed into viable opportunities, using the traditional brainstorming approach of building on others' ideas, extending and adapting, polishing and improving, until something valuable emerges.
- *Good luck.* In this version of the brainstorm or interview, people imagine what might happen if they were the luckiest person on earth. Some of the

resulting ideas will be crazy and unrealistic, but some will be worth exploring further and could produce genuine opportunities for the project.
- *In your dreams.* For many people, the future of their project appears to be a nightmare waiting to happen. Participants in a dream workshop or interview are invited to imagine a wonderful dream in which everything on their project is going well, then to describe the circumstances that produce that outcome in the dream. If you dreamed about your project and woke up smiling, what would you be dreaming?
- *What if not?* Chapter 2 talked about opportunities that might arise from the inverse of threats or the absence of threats, as well as those that result from responses to threats and others that are independent of threats. A "What if not?" workshop or interview starts with a list of threats and asks people to imagine their opposites, as well as considering whether any positive uncertainties might emerge if a threat didn't happen.

It is also possible to modify other forward-looking techniques, including *futures thinking, scenario analysis,* and *visualisation.* A specific technique that combines aspects of all three of these approaches is the pre-mortem.

- *Pre-mortems.* A medical post-mortem is conducted after death to determine the cause. In 1998, Gary Klein[*] developed the project pre-mortem technique, also known as prospective hindsight. This involves starting at the project end-point and imagining the paths that could have led to this outcome. While this technique is often used to explore the conditions that might lead to project failure, it is of course also possible to consider how we might end up with a successful outcome. Having plotted potential paths to success, we can take those insights into the planning for the future of our project. While this doesn't immediately lead to identification of specific opportunities, it can encourage the positive mindset that is required to start looking for upside risks.

## USING "TWO-DIMENSIONAL TECHNIQUES"

Rather than try to modify existing threat-focused risk identification techniques so that they can also be used to find opportunities, it might be better instead to use *techniques that explicitly look for both threats and opportunities together.* We might call these *"two-dimensional techniques"*, as they have a dual focus on both types of risk.

---

[*] (Klein, 1998, 2007)

Two such techniques are described below (although others are available):

- SWOT Analysis (Strengths, Weaknesses, Opportunities, Threats)
- Force-Field Analysis

## SWOT Analysis

SWOT Analysis is an obvious candidate technique for explicit identification of both opportunities and threats. It is commonly used to support strategic decision-making, but it can easily be modified to identify risks at project level. When used in this way, its four elements are:

- *Strength*—an existing characteristic, resource, or capacity of the project organisation that helps the project to achieve its objectives
- *Weakness*—a limitation, fault, or defect in the project organisation that hinders the project from achieving its objectives
- *Opportunity*—an uncertainty that, if it occurred, would have a positive effect on achievement of project objectives (upside risk)
- *Threat*—an uncertainty that, if it occurred, would have a negative effect on achievement of project objectives (downside risk)

Strengths and Weaknesses are *facts about the project organisation*, answering the question "Who are we?" Opportunities and Threats are *uncertainties relating to project objectives*, and are about "What are we doing?" It is obvious that "Who we are" affects "What we do", and so SWOT Analysis at project level is based on the relationship between Strengths/Weaknesses and Opportunities/Threats. Typically, strengths lead to opportunities, and weaknesses produce threats, although this is not exclusively the case.

SWOT Analysis is best undertaken in a facilitated workshop setting, attended by project team members with appropriate knowledge and expertise. The workshop has two sections, as illustrated in Figure 5-3:

1. Identify strengths and weaknesses using brainstorming.
2. Identify opportunities and threats that might arise from these strengths and weaknesses.

The first element of the workshop is to undertake a structured brainstorm that identifies existing strengths and weaknesses of the project organisation. This should include generic characteristics, as well as factors relating to the particular project under consideration. Example strengths might be the organisation's

**92** Capturing Upside Risk: Finding and Managing Opportunities in Projects

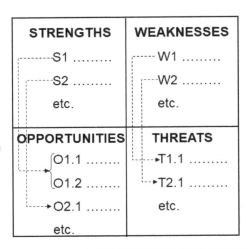

Step 1: Identify strengths and weaknesses using brainstorming

Step 2: Derive opportunities from strengths, and threats from weaknesses, using risk metalanguage

**Figure 5-3**  SWOT Analysis Process

position as market leader, high levels of technical expertise, or relevant project team experience. Weaknesses could include poor subcontract management, low project team morale, or a tendency to change project priorities.

The workshop then considers each strength and weakness in turn to identify related opportunities and threats. This can take the form of simple "So what?" questions: "Yes, we're the market leader, but how might that help this project? Given our technical expertise, what might happen that helps us deliver this project early? We know we have poor subcontract management processes, but so what?"

As an alternative to "So what?" questions, risk metalanguage (discussed above) can provide a more structured framework for turning strengths and weaknesses into opportunities and threats:

"As a result of <strength/weakness>,
　　<uncertain event> may occur,
　　　　which would lead to <one or more effects on objective(s)>."

It is common for each strength and each weakness to give rise to more than one risk, which can provide useful information on common causes of risk (as illustrated in Figure 5-3).

Example risks are shown in Table 5-4, using the metalanguage format.

The key advantage of SWOT Analysis is that it explicitly identifies both opportunities and threats using a single technique. It also naturally deals with opportunities before threats, ensuring that these are given proper attention. The

Table 5-4 Example Risks Derived from SWOT Analysis

| Example Opportunities from Strength | Example Threats from Weakness |
|---|---|
| S4 We have an existing supply chain. | W2 Our experience is in a different segment from the current client. |
| O4.1 Because of S4, we may get preferential rates from one or more suppliers, leading to cost savings. | T2.1 Because of W2, we may misunderstand the client's needs/wants, resulting in failure of user acceptance tests. |
| O4.2 Because of S4, fewer problems may occur during interface testing, creating additional float in the schedule. | T2.2 Because of W2, unexpected regulatory requirements may exist, leading to rework to meet the regulations. |

downside of SWOT Analysis is its focus on internally generated risks arising from the project organisation itself, from the effect of "who we are" on "what we're doing". As a result, SWOT Analysis cannot find external risks that arise from the project environment or context. This means that it must be used in conjunction with other risk identification techniques that don't have the same limitation.

## Force-Field Analysis

Force-Field Analysis was developed by Kurt Lewin in 1951[*] as a creative process to identify the *driving and restraining forces* of social change:

- *Driving forces* initiate and sustain change.
- *Restraining forces* hinder change and keep the situation at its current level.

This technique is now widely used in change management and strategic decision-making to identify positive and negative influences on achievement of objectives. It is simple to adapt this approach for identifying project risks, either analysing the forces that affect achievement of each project objective, or considering the whole project.

Force-Field Analysis has the following steps:

- Simply and clearly state the project objective being analysed.
- List all forces or influences which support or drive achievement of the objective, and those forces or influences which oppose or restrain change.

---

[*] (Lewin, 1951)

- Rate the strength of each force, using a scale from 1 (Weak) to 5 (Strong), and draw a force-field diagram for the objective, with the size of each force reflecting its strength.
- Assess the current situation to determine the balance of existing forces.
- Consider what risks might affect the various forces:
  - *Threats* are uncertainties that would either strengthen a restraining force or reduce a driving force, making it harder to achieve the project objective.
  - *Opportunities* are uncertainties that would either weaken a restraining force or strengthen a driving force, making it easier to achieve the project objective.
- Estimate the effect of each risk on the driving and restraining forces.
- Reassess the balance of forces in relation to the selected objective, taking account of identified risks.

An example simple Force-Field Analysis diagram is shown in Figure 5-4 for a project that aims to introduce new production hardware. For this objective, threats would include any uncertainty that might reduce customer demand,

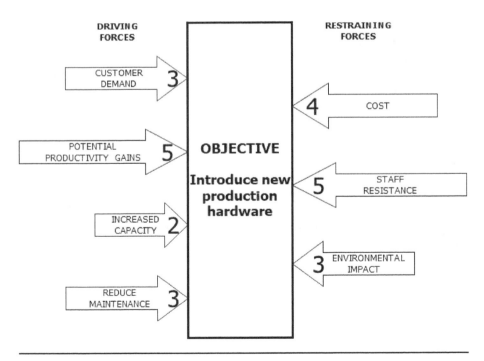

**Figure 5-4** Example Force-Field Analysis Diagram

lead to lower productivity gains, or increase cost. Example opportunities include uncertainties that might lead to lower staff resistance, less environmental impact, or make maintenance easier/cheaper. The challenge for people using this technique is to be specific about the threats and opportunities that might affect the balance of forces. The advantage is that Force-Field Analysis naturally leads to identification of both threats and opportunities as part of the same analytical process.

## USING RISK METALANGUAGE TO IDENTIFY OPPORTUNITIES

Using risk metalanguage as a structure for risk descriptions encourages us to separate the three elements of cause/risk/effect in our thinking. This can be particularly helpful when we come to identify opportunities, because we can work from either end of the cause/risk/effect chain into the middle to find risks, as illustrated in Figure 5-5. Risk metalanguage uses the format:

"As a result of <one or more definite causes>,
    <uncertain event> may occur,
        which would lead to <one or more effects on objective(s)>."

We can start with positive causes and see how they might produce positive uncertainties, or we can start with positive effects and consider how they might come to pass.

**Figure 5-5** Using Risk Metalanguage to Identify Opportunities

This technique starts with either a positive fact that might give rise to opportunities (cause), or with a positive impact on a project objective (effect). Then simple questions are used to discover opportunities:

- Starting from the cause: "This positive fact is true about our project, but *so what?* What opportunities might result from this fact? How could it help us? What positive uncertainties does it introduce into our project?"
- Starting from the effect: "This positive outcome might be theoretically possible, but *how and why might it happen?* What uncertain event or circumstance that we're not currently expecting would lead to this positive result if it happened? How might we be lucky enough to end up here?"

If we're not used to identifying opportunities, this approach can be particularly powerful. We can think of good things that are true about our project (this is a high-profile top-priority project, the team are excited to be working on it, etc.), then use those as a starting point to find opportunities. We can also easily imagine positive outcomes (we might deliver early or come in under budget, the company's reputation might be significantly enhanced, etc.), and then think about what opportunities might lead us there.

## SEPARATE OR TOGETHER?

When people realise that the risk identification step needs to identify both threats and opportunities, they commonly ask whether it is better to look for both types of risk in a single exercise, or to treat them separately. This particularly affects techniques that involve facilitated workshops, such as brainstorming, although it can also apply to the use of structured interviews or questionnaires. There is no single best answer to this question, and each approach has its benefits and shortcomings.

- *Together.* Using a single risk identification workshop or interview to find threats and opportunities together can appear to allow more efficient use of time and effort, as well as giving participants the freedom to raise any kind of risk whenever they think of it. But it's easy to get over-focused on looking just for threats and fail to give enough consideration to identifying opportunities. Effective facilitation can overcome this tendency, reminding people to give equal attention to both types of risk.
- *Separate.* Maybe it's better to divide our workshop or interview into two sections, one dealing with threats and another with opportunities. This has the advantage of focus, allowing people to concentrate on one type of risk at a time. However, it is more restrictive for participants, preventing

them from raising risks as they think of them, and forcing them to remember some risks until later. Also, if we separate risk identification into two sessions, it raises the question of which to do first.
- Some people think we should address opportunities first, while participants are fresh and full of energy, because they are less used to thinking about upside risk and this might require more effort. Leaving threats until later is OK because it's easier to think about them.
- Others suggest that since people naturally think first about threats, we should get those out of the way before turning our attention to opportunities. The danger here is that we never get around to identifying opportunities, and instead we spend all our time and energy in the first session on threats.

There's no right answer to this question, and it may depend on the preference of the facilitator or workshop participants, or on the culture of the organisation or project team. But if we decide to address threats and opportunities in separate risk identification exercises, we need to agree which way to do it, then stick to it, to ensure that we do actually spend adequate time and effort looking for both types of risk.

## A MINDSET FOR OPPORTUNITY

The preceding sections have identified a number of effective techniques that we can use to identify opportunities. Some of these are modifications of approaches that are commonly used to find threats, including:

- Constraints analysis (similar to assumptions analysis)
- Benefit tree analysis (based on fault tree analysis)
- Upside brainstorms and interviews (three wishes, good luck, dreams, and "What if not?")
- Pre-mortem (combining futures thinking, scenario analysis, and visualisation)

Other techniques are specially designed to find both threats and opportunities in the same approach, for example:

- SWOT Analysis
- Force-Field Analysis

We have also discussed whether it is better to adopt an integrated approach to identifying both threats and opportunities together or aim to identify opportunities in a separate exercise. However, the key to successfully identifying opportunities is not the technique we use or the format of the risk identification session. Risks (including opportunities) are not identified by techniques; they are identified by people. And one of the main barriers to identifying opportunities is *the mindset of the people who are looking for risks.*

A negative mindset that believes all risks are bad will drive us to seek only threats. This needs to be challenged and changed if we are to stand any chance of finding the real opportunities that are available in our projects.

A mindset is *a set of attitudes and beliefs that drive behaviour in a particular direction.* Mindset is useful because it creates a sense of focus, allowing us to concentrate on the things that we believe are important. But most people are not aware of their mindset, since it appears only natural to them to view things in a particular way. They need to be educated and persuaded to change their mindset, and this takes time and sensitivity. You can't command someone to change their mind!

A mindset for opportunity has a number of defining characteristics:

- *Realism.* Recognise that the world is not entirely hostile, but it includes friendly factors and forces. There's no reason why you should be particularly unlucky or exposed only to adverse circumstances and influences.
- *Think positive.* Good things happen (as well as bad), and we can improve our chances of being "lucky" if we're aware of what those good things might be.
- *Be alert.* Look out for opportunities in every circumstance. Don't miss anything that could be helpful to you.
- *Be curious.* Examine unfamiliar situations, explore, play, ask questions. Who knows what opportunities may be hiding in plain sight, or discoverable with a little effort?
- *Be ready.* Expect the unexpected. Plan for a little serendipity or coincidence. Include slack time in your schedule for the nice surprises that will come your way.
- *"Can do" attitude.* Don't view yourself as a victim of circumstances, but take responsibility for the way you respond to the situation you find yourself in. Have clear objectives and be determined to use your best efforts to achieve them. This needs to be tempered with realism, not setting unachievable goals, and not overestimating your abilities and competence. Be aware of your strengths (and use them) and your weaknesses (and address them).
- *Visualise success.* Imagine good outcomes, positive results, successful achievements, then plan backwards to work out how to reach them.

Some people think like this naturally, but for many of us it is unfamiliar territory, perhaps due to our culture, our upbringing, our education, or our previous experience. If we truly believe that opportunities exist that could make a significant difference to our project, and that these opportunities can be found and captured, then we'll make the effort to change our thinking and behaviour. This will require application of *emotional intelligence*,[*] seeking to understand ourselves, being self-aware of our underlying attitudes and drivers, and taking active and intentional steps to *modify our mindset*. A skilled and sympathetic mentor or coach can be very helpful, holding up a mirror to our internal world, offering guidance on effective ways to change, and providing feedback as we make progress.

Mindset matters when it comes to finding opportunities. There's no doubt that most people find it hard to identify opportunities when they first try, and this is largely driven by their belief that all risks are bad. But focused attention to the way we think about risk will help us to modify our mindset. As we start to think differently about risk, our behaviour will follow. When this is supported by effective techniques and skilled facilitation, identifying opportunities will become easier.

## WRITE IT DOWN

The risk identification step has the vital role of ensuring that we find the real risks that could affect achievement of our project objectives, if they happened, including both threats and opportunities. But identifying them is not enough. Having spent time and effort discovering these uncertainties that matter, we need to record our findings in a way that can be used later in the risk process. This is the purpose of the risk register—to record data on identified risks in a consistent and useful format.[†]

The style of risk register required for this project will be defined in the Risk Management Plan (RMP, see Chapter 4), as part of tailoring the risk process to the risk challenge of our project.

The risk register needs to be initiated the first time we perform risk identification, because it will form the input to the prioritisation and analysis stages that follow. It also acts as an auditable record of our findings, and it can be used retrospectively as part of our lessons-learned exercises. We'll add more details on each risk as the risk process continues, including prioritisation information, risk owner, response details and status, etc. At this stage in the risk process, we launch the risk register by completing the header information section for our project (if this is present in our chosen risk register format), as well as recording risk identification data for each identified risk, as shown in Table 5-5.

---

[*] (Hillson & Murray-Webster, 2007)
[†] (Williams, 1994)

Table 5-5 Risk Data to Record after Risk Identification

| Data Field | Minimum | Typical | Detailed |
|---|---|---|---|
| **Header Information** | | | |
| Project data (project title, project manager, client, project status) | | X | X |
| Document data (Risk Register issue number, date, approvals) | | X | X |
| Date of most recent risk review | | X | X |
| Date of next risk review | | | X |
| **Risk Identification Data** | | | |
| Unique risk identifier | X | X | X |
| Date identified | | X | X |
| Name of person identifying risk | | | X |
| Threat/Opportunity indicator | | | X |
| Short risk title | X | X | X |
| Full risk description (cause/risk/effect) | X | X | X |
| Explanatory comments | | | X |
| Risk status (for example draft, active, closed, occurred etc.) | | X | X |

When we consider recording both threats and opportunities in the risk register, we encounter the "separate or together" dilemma again. Should we have two risk registers: one for threats and another for opportunities? The answer is in the name "risk register". The risk register registers risks. The term "risk" includes both threats and opportunities, so the risk register must also include both. It's helpful to have a field in the risk register that indicates whether the risk is a threat or an opportunity, although this should be evident from the risk description and impact ratings, which will both be positive. But if we have a flag to show which type of risk it is, we can then use filters to separate opportunities from threats if this is necessary.

We'll come across this issue again in the next step of the risk process when we consider how to prioritise risks: Should we have two lists of "top threats" and "top opportunities", or one overall "top risk" list? Chapter 6 reveals the answer!

## SUMMARY AND REFLECTION QUESTIONS

The only risks we can manage proactively are the ones we can identify in advance. We're used to this thinking when we consider threats, and much of our risk management effort is focused on preventing threats from occurring or protecting ourselves from their potential effects. But the same is true of opportunities. An unidentified opportunity cannot be captured or used to help the project achieve its objectives. If we're serious about including opportunities in our risk management approach, we must first be able to identify them.

Fortunately, several commonly used risk identification techniques can be used to find opportunities, even though we usually focus on threats. In addition to these, there are specific techniques that are designed to find both threats and opportunities in the same process. Using these "two-dimensional techniques" reminds us that we should be looking for both types of risk, not just threats.

Although techniques exist to help us find opportunities, the biggest barrier is often ourselves, because threats have dominated our previous experience, habits, actions, and thinking. The risk identification step is the first place in the risk process where we need to challenge this mindset, actively training ourselves to seek out opportunities that could help us to achieve our project objectives. Only then can we gain the benefits that opportunities offer our project, maximising our chances of project success.

### Reflection questions

- What is the difference between "business-as-usual risks" and "real risks"?
- Why is it not possible to identify all risks at any particular point in time? What can you do about it?
- Name one risk identification technique from each of the three time-perspectives—past, present, future—and explain how each one might be used to identify opportunities.
- Explain risk metalanguage and describe how it can be used to identify opportunities.
- What aspects of your mindset make it hard for you to identify opportunities? How might you overcome these?

## THE NEXT STEP ("NOW WE'VE IDENTIFIED THE OPPORTUNITIES, WHICH ONES ARE MOST IMPORTANT?")

Risk identification is only the starting point of our journey towards managed risk. Ideally it results in a list of the real risks to the project, described clearly and unambiguously, including both threats and opportunities.

But not all identified risks are equally important. In order to avoid wasting time on insignificant or irrelevant risks, we need to pick out the ones that matter most, including the worst threats and the best opportunities. This requires a robust framework for prioritising risks, which leads us to the next step in the risk management process—qualitative risk assessment.

# Chapter 6
# Picking Winners

Now we know how to find opportunities, but they are not all worth pursuing. Some will be crazy ideas that are extremely unlikely to happen, or wishful thinking dreams that belong in the realms of fantasy. We need to look through the opportunities we've identified and pick out the best ones—opportunities that we can chase and capture, turning them into tangible savings or benefits to help us achieve our project goals.

This chapter explains the basic principles of qualitative risk assessment, focusing first on how we prioritise threats. Fortunately, all the commonly used threat-based assessment techniques that work for threats also work for opportunities! Applying the same approach to opportunities that we usually use for prioritising threats, we can find the opportunities that are worth further attention and action, **picking the winners** that will have the biggest positive effect on our project.

## PURPOSE AND PRINCIPLES OF QUALITATIVE RISK ASSESSMENT

Risk identification done properly results in a list of risks which can be quite long, including both threats and opportunities, recorded in a risk register. This is a good start to the risk process, but it gives us an immediate problem. We nearly always identify more risks than we can manage, given the amount of time and effort available. Also, it stands to reason that some risks will be more important than others. Some will require urgent focused attention and action, while it may be safe to leave some risks until later, or perhaps even ignore them altogether.

In order to decide which threats and opportunities we should focus on, it is essential that we have a reliable way of *prioritising risks for further attention and action*. We need to produce a ranked risk list that tells us and our project stakeholders which risks are currently the most important, that have the greatest potential to affect achievement of our project objectives, either positively or negatively. These are the threats and opportunities that demand our *attention*. We also need to determine which risks require robust or radical responses, and when and where to apply our necessarily limited risk resources—these are the threats and opportunities on which we need to take *action*.

In addition to finding out which are the most important individual threats and opportunities, it's helpful to consider whether there are any significant links between risks that we need to know about and act on. Grouping risks by various shared characteristics can help us to develop generic risk responses that effectively address several risks at once.

This is what the next step of the risk process is all about. ***The purpose of qualitative risk assessment is to evaluate key characteristics of identified risks in order to prioritise them for further attention and action.***

As we work out how best to assess our risks to meet this goal, there are several **principles** to bear in mind.

## Use consistent and objective assessment framework

When people consider how important each risk is, the result is always subjective: "I think this risk is extremely important." "Oh really? I'm just not bothered about this one."

There are very few risks about which we actually know all of the relevant characteristics in sufficient detail to be able to decide how important the risk is. If we want to prioritise risks, then we need to assess them all on the same basis, using a *consistent framework*.

We should also seek to define our terms in such a way that we *minimise subjectivity* and provide a basis for discussion. This requires us to develop a language of assessment that everyone can use, supported by common terminology and definitions. Shared language will not guarantee agreement, of course, but it will allow us to have meaningful conversations about why we might disagree when assessing a risk, how far apart we are in our assessments, and where we need to obtain additional information to help us reach an agreed assessment. For example, if we all agree what we mean by "high schedule impact" and "low schedule impact", and if these terms are quantified (e.g., "high" means three to six months change in delivery date, "low" means under one month), then the debate is not about whether the impact assessment should be "high" or "low" (which is entirely subjective and largely unproductive), but whether we believe that if this risk occurred it would affect the delivery date by a few weeks or a few months. Perhaps we can present hard evidence to support one view or the other, or quote relevant previous experience, enabling us to reach consensus on the basis of objective data.

## Reflect corporate risk appetite and risk thresholds

When we come to create an agreed framework that defines what we mean by High/Medium/Low for each dimension of the risk assessment, we should not produce definitions in a vacuum. The specific risk assessment criteria for this project should be based on a clear understanding of corporate risk thresholds which express the overall risk appetite of senior leadership. These top-level risk thresholds should have been broken down across the various levels of the business, through departments and functions, via portfolios and programmes to project level, ultimately allowing risk assessment criteria to be generated for this project. This process is described in Chapter 4, and the results are documented in the Risk Management Plan (RMP) for this project.

If we don't have this context, then any risk assessment framework that we produce for our project will be based on a guess about how much risk is too much risk.

## Seek input from different perspectives

Even if we have an agreed risk assessment framework that reflects corporate risk thresholds, we will still have to place each identified threat and opportunity within the framework. It is very unlikely that any one person will have the necessary experience or expertise to assess every risk properly—not even the project manager. As a result, we should involve others in the qualitative risk assessment step wherever possible. This should include project team members, of course, especially those with particular role expertise (technical leads, procurement staff, planners/estimators, etc.). It would be valuable to include the project sponsor to represent the business and client perspectives. For some projects it might be appropriate to invite suppliers or contractors or user representatives to contribute. (This use of multiple perspectives is also important in the preceding risk identification step, as discussed in Chapter 5.)

As we broaden the range of perspectives, the assessment challenge becomes more difficult, as each party comes with a set of drivers and goals, some of which will be incompatible or conflicting. A skilled facilitator can help a group of people to reach a good outcome with an agreed realistic assessment of how important each risk really is to the success of the project.

## Beware bias

Everyone brings a set of filters to work, including *cognitive biases* and *subconscious heuristics*.[*] These influence our perception and behaviour in complex ways that we often don't understand, leading us to view some risks as more important than they really are, and to discount others that might be very significant. This can have a big effect in the qualitative risk assessment stage, because assessment is necessarily subjective. We rely on gut feel and instinct to inform our judgements, and bias can be hard to spot, especially when we are considering things that are uncertain by nature. When we lack hard evidence, we naturally fall back on biases and heuristics to help us make a decision.

While we may be largely unaware of our own sources of bias, we can often see them in action in others. A skilled facilitator can therefore help each one of us to recognise when we are being influenced by unconscious drivers, which we can then counter.

## Summary

The purpose and principles of qualitative risk assessment are summarised in Table 6-1.

Table 6-1 Purpose and Principles of Qualitative Risk Assessment

| Purpose | To evaluate key characteristics of identified risks in order to prioritise them for further attention and action |
|---|---|
| Principles | • Use consistent and objective assessment framework<br>• Reflect corporate risk appetite and risk thresholds<br>• Seek input from different perspectives<br>• Beware bias |

# DEFINING PRIORITISATION DIMENSIONS

Qualitative risk assessment allows us to prioritise identified threats and opportunities, and we need to rank risks for two important reasons:

- *Attention*—which risks demand our focus now, which ones do we need to know about, understand, and consider?
- *Action*—which risks need to be addressed now in order to position and prepare ourselves most effectively to deal with them?

---

[*] (Kahneman, Slovic, & Tversky, 1986; Slovic, 1987; Tversky & Kahneman, 1974)

The risks requiring our **attention** are obviously the ones that are most important. But there are several reasons that we might consider a risk to be important. Size is a key factor, of course. Risk is "uncertainty that matters", and the size of each threat or opportunity depends on how we answer two related questions: "How uncertain? How much does it matter?" The simplest answers to these questions are found if we consider the *probability* that a one-off risk will occur (or its *frequency* if it is a repeated event), as well as the potential *impact on objectives* if it does occur.

There are other reasons that we might need to pay attention to a risk, in addition to its size. Other characteristics of a threat or opportunity can make it important—for example, its potential to affect strategic goals or objectives outside the scope of the project (*strategic impact*), its influence on other risks (*connectedness*), or the importance attached to a risk by key stakeholders (*propinquity*). These terms are defined in Table 6-2.

As well as prioritising risks so that we can pay attention to the most important threats and opportunities, we also need to identify the risks where **action** is required. We will clearly need to deal with the big risks, as determined by their probability and impact. But there are other factors to consider when prioritising threats and opportunities for action, including how soon the risk might be expected to occur (*proximity*), how soon we must act in order to manage the risk effectively (*urgency*), and the degree to which we are able to influence the occurrence of a risk (*manageability*) or the extent of its impact (*controllability*). These terms are explained in Table 6-2.

If we are going to use any of these prioritisation factors to rank threats and opportunities for attention or action, we will need defined scales against which we can assess each identified risk. These scales are provided in the RMP for probability, frequency, and impact (see Table 4-4). Where the RMP specifies that other risk characteristics will be used to prioritise risks, scales for each of these factors should also be included in the RMP, defining what is meant on this project for Low/Medium/High in each of the relevant dimensions. Table 6-2 gives some examples of how each parameter might be defined.

## TYPICAL TECHNIQUES FOR PRIORITISING THREATS

Risk prioritisation is generally based on assessment of two characteristics of identified risks: probability/frequency and impact. It is possible to consider a range of other parameters in addition, as outlined above. Where a project is only using two dimensions for prioritisation, a simple two-dimensional matrix works well as a ranking tool, and this is the purpose of the Probability-Impact Matrix (P-I Matrix). However, if the RMP specifies a more rigorous prioritisation

Table 6-2 Risk Prioritisation Factors

| Risk Characteristic | Description | Examples |
|---|---|---|
| **General** | | |
| Probability | The level of confidence of the assessor that this (one-off) risk will occur | Very Low<br>1–5% confident that this risk will occur. |
| Frequency | The frequency with which this (repeatable) risk is expected to occur | Medium<br>Once in six months. |
| Impact | The estimated effect of this risk, if it occurs, on one or more project objectives (negative for threats, positive for opportunities) | High<br>1–3 months change in total float on critical path.<br>5–10% change in total project cost. |
| Risk score | A numerical representation of the combination of probability/frequency and impact | 0.12, representing Low probability/High impact. |
| **Prioritising for Attention** | | |
| Strategic impact | The estimated effect of this risk, if it occurs, on one or more strategic or programme-level objectives (negative for threats, positive for opportunities) | Low<br>Minor effect on one or two strategic or programme objectives outside the project. |
| Connectedness | The extent to which occurrence of this risk might be correlated with the probability or impact of one or more other project risks | Medium<br>Strong correlation between this risk and 10–20 other risks. |
| Propinquity | The extent to which this risk is perceived to matter to one or more stakeholders | Very High<br>One stakeholder regards occurrence of this risk as a showstopper. |
| **Prioritising for Action** | | |
| Proximity | How soon this risk is expected to occur, if it does occur ("impact window") | High<br>This risk is expected to occur in the next 1–2 months. |
| Urgency | How soon responses must be implemented in order to address this risk effectively ("action window") | Very High<br>Risk responses must be implemented next week in order to be effective. |
| Manageability | The extent to which planned risk responses are able to influence the occurrence or impact of this risk before it has occurred | Very Low<br>Risk occurrence is driven by outside factors that cannot be influenced. |
| Controllability | The degree to which the risk owner can influence the impact of this risk after it has occurred | High<br>Good fallback plans are in place to ensure that any impact is brought within acceptable levels. |

approach using more than two factors, the matrix approach is not suitable, and other tools are required. These are discussed in the following sections.

## Two dimensions: P-I Matrix

The purpose and format of the two-dimensional P-I Matrix for our project are defined in the RMP and discussed in Chapter 4, with an illustration in Figure 4-4. When using the P-I Matrix to prioritise threats, each identified threat is assessed in two dimensions:

- Probability (level of confidence that the risk will occur), or frequency (chance of occurrence in a given time period).
- Impact (effect on one or more project objectives). The impact of threats is negative, of course.

The RMP also defines different levels of probability/frequency and impact to be used when assessing risks in our project, telling us precisely what is meant by High, Medium, Low, etc. (see Chapter 4, with an example given in Table 4-4).

Assessing these two parameters allows the threat to be placed in one of the cells on the P-I Matrix, and it can then be ranked using the prioritisation zones (Red = top-priority; Yellow = medium-priority; Green = low-priority).

One immediate question arises when considering how to implement this approach. If a risk can affect more than one objective, and the extent of that effect might be different against different objectives, how can this be plotted on the P-I Matrix? As an example, consider Risk X with the following assessment:

- Probability = Medium
- Impact against schedule = High
- Impact against budget = Low
- Impact against performance = Nil

At first sight it is hard to see where this risk belongs in the P-I Matrix. Several solutions are possible:

- Use a separate P-I Matrix for risks affecting schedule, budget, or performance, and plot the risk on each P-I Matrix in the appropriate cell. Risk X would then appear as Medium/High on the schedule P-I Matrix, and as Medium/Low on the budget P-I Matrix, but it would not be plotted on the performance P-I Matrix. This approach has the advantage of bringing together all the risks that can affect each project objective, allowing an

110  Capturing Upside Risk: Finding and Managing Opportunities in Projects

overall assessment of the level of risk associated with each objective. On the downside, it is more complex and requires more effort to plot each risk multiple times.
- Use a single P-I Matrix, and plot the risk more than once, using a code to indicate which type of impact is being represented. For Risk X, we could plot $X_S$ as Medium/High on the P-I Matrix, and $X_B$ as Medium/Low, showing both the schedule and budget impacts on the same grid. The problem with this approach is that the P-I Matrix quickly gets very full, with each risk appearing multiple times, making it hard to see what is going on.
- Use a single matrix and apply an algorithm to combine the impact of a risk against various dimensions. While this sounds promising as an approach, it is hard to design a suitable algorithm that preserves all the information. We can't simply average out the impact types (for Risk X, schedule impact is High, budget impact is Low, so we treat it as an overall Medium impact) because this can underplay the importance of a risk, especially if it has a High or Very High impact in just one dimension.
- Use a single P-I Matrix, taking only the highest-ranking impact type to determine where the risk is plotted. Risk X then plots as Medium/High, based on the High schedule impact. This has the advantage of simplicity,

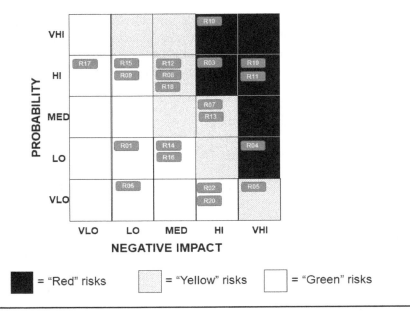

Figure 6-1  Plotting Threats on the P-I Matrix

and rightly treats all impact types as equivalent (a High impact on schedule is as important as a High impact on budget). On the downside, some of the assessment information is not used for prioritisation.

On balance, this last approach is generally accepted as the best way to treat risks with impacts against more than one objective.

Taking this approach, we can plot each threat on a single P-I Matrix and determine which are top-priority (in the Red zone), which are medium-priority (Yellow), and which are low-priority (Green), as shown in Figure 6-1.

This level of prioritisation allows us to divide our list of identified threats into three groups: top/medium/low priority. In some cases, this may be sufficient for us to decide which risks require attention and action. For example, we may decide to focus risk response development on top-priority (Red) risks, to monitor medium-priority (Yellow) risks, and to take no action on low-priority (Green) risks.

## More detailed prioritisation in the P-I Matrix

The P-I Matrix treats all risks in a Red/Yellow/Green prioritisation zone as equally important (all Red risks are equally red), which might be OK for some projects. On other projects, we may wish to take prioritisation to another level of detail, especially if there are a large number of risks within each zone. This means developing a way of ranking risks within each zone, to pick out the reddest risks in the Red zone, for example. This is done by using a *risk scoring system*, with numerical scores used to represent each level of probability/frequency and impact. Then we multiply the probability and impact scores together to give an overall risk score for each risk.

The simplest version of a risk scoring scheme is to use a linear sequence of whole numbers for each scale point, and to use the same scores for each dimension. For example, if we are using a three-point scale for probability and impact, we could replace Low with 1, Medium with 2, and High with 3, for both probability and impact. A five-point scale might represent Very Low/Low/Medium/High/Very High as 1/2/3/4/5. This latter would result in the set of risk scores in Figure 6-2.

This approach is commonly used, with prioritisation zones defined using the resulting scores. For example, we might define the following zones in Figure 6-2:

- Red    > 12
- Yellow  5–12
- Green   < 5

112  Capturing Upside Risk: Finding and Managing Opportunities in Projects

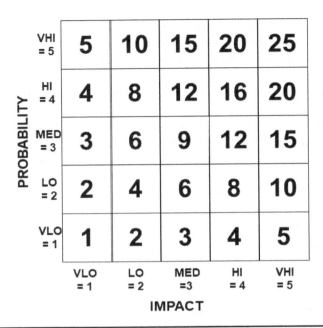

**Figure 6-2**  Risk Scoring Using Two Linear Sequences 1/2/3/4/5

Although this scoring scheme is popular, it has several major disadvantages. Firstly, several cells have the same score, making it impossible to prioritise risks uniquely. More importantly, since the two scales are identical, the resulting prioritisation grid is symmetrical. For example, if Risk A has High probability/ Low impact, and Risk B has Low probability/High impact, a symmetrical grid prioritises these equally (both scoring 8 in Figure 6-2). But our earlier discussion explained that intuitively we feel we should weight impact higher than probability; in other words it feels right that Risk B (a small chance of a major impact) should be higher priority than Risk A (quite likely to happen but with only a minor effect). To achieve this weighting of impact over probability, it is necessary to use a non-linear set of scores in the impact dimension, so that risks with a High or Very High impact receive a higher risk score than risks with Low or Very Low impact, even if the probability is above Medium.

One risk scoring system that embodies this feature is shown in Table 6-3. This has a linear scale for the probability/frequency dimension, and a non-linear scale for impact. In this example, both scales are set between 0 and 1, but this is not essential. When the scales in Table 6-3 are multiplied together, the resulting risk scores are as illustrated in Figure 6-3.

In contrast to the risk scores in Figure 6-2, we can see from Figure 6-3 that the score in each cell of the P-I Matrix is unique. This means that no two combinations of probability/impact result in the same score. We can also see the

Table 6-3  Example Risk Scoring Scheme with Non-Linear Impact Scale

| Scale | Probability/ Frequency Score (Linear) | Impact Score (Non-Linear) |
|---|---|---|
| VLO | 0.1 | 0.05 |
| LO | 0.3 | 0.1 |
| MED | 0.5 | 0.2 |
| HI | 0.7 | 0.4 |
| VHI | 0.9 | 0.8 |

effect of the non-linear impact scale, making the prioritisation grid asymmetrical, skewed towards impact. As a result, with the two risks mentioned above (Risk A with High probability/Low impact, and Risk B with Low probability/High impact), the non-linear impact scale means that Risk B (score 0.12) is rightly prioritised above Risk A (score 0.07).

We can also use these risk scores to produce asymmetrical prioritisation zones in the P-I Matrix. For example, the zones in Figure 6-1 are defined as follows:

- Red      > 0.2
- Yellow   0.08–0.20
- Green    < 0.08

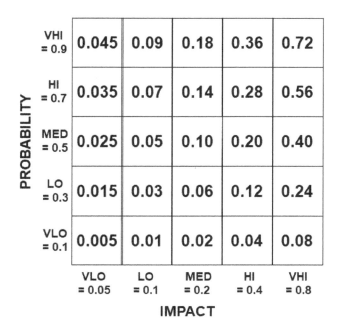

Figure 6-3  Risk Scoring Using Linear/Non-Linear Sequences

Having a set of unique risk scores allows us to prioritise risks unambiguously, and it provides more detail than simply dividing risks into three priority groups (Red, Yellow, Green). With this type of risk scoring, we can determine which risks in the Red zone are "most red", for example. We can produce a prioritised list of "top risks", based on the risk scores.

We might also devise various Key Risk Indicators (KRIs) based on risk scores, such as Total Risk Score for active risks (TRS), Average Risk Score for active risks (ARS), or Relative Risk Exposure Index (defined as the ratio of the current value of [TRS * ARS] compared with a baseline value). The index is 1.0 if the current level of risk exposure is the same as the baseline, with less than one indicating reduced risk exposure compared to the baseline, and higher than one meaning increased risk exposure over the baseline.

## Three dimensions: Other approaches

The P-I Matrix is an ideal way to prioritise threats if we are only considering the two dimensions of probability/frequency and impact. But we've seen earlier in this chapter that the importance of a risk does not only depend on how likely it is to occur or what its potential effect on achievement of objectives might be. Table 6-2 lists ten prioritisation criteria that we might use to determine which risks are the "most important". Some of these characteristics are useful if we want to know which risks demand our *attention* now, and others are more relevant if we need to decide where to take *action*.

If we want to use more than two prioritisation criteria, we can't use a two-dimensional matrix to rank risks. Instead, we need different tools that can handle more factors. The more characteristics we consider, the more complex the analysis becomes, so it is usual to limit this to just three parameters, but we can of course choose various combinations of criteria. It is also possible to use the risk score (representing the combination of probability/frequency and impact) as a single composite prioritisation factor, effectively allowing us to consider four distinct parameters.

Two of the more common multi-criteria prioritisation tools are the bubble chart and the risk prioritisation chart, illustrated in Figures 6-4 and 6-5.

The bubble chart shows three parameters, using the $x$ and $y$ axes as well as the size of the bubble. We need to think carefully about which parameters to choose, as well as how to interpret the results. The example in Figure 6-4 plots *urgency* (how quickly we need to act) on the *x*-axis, *manageability* (how easy the risk is to manage in advance) on the *y*-axis, and *severity of impact* is reflected by bubble size. In this case, a small bubble (low impact) is acceptable anywhere on the chart and does not require immediate action, even if it is urgent and

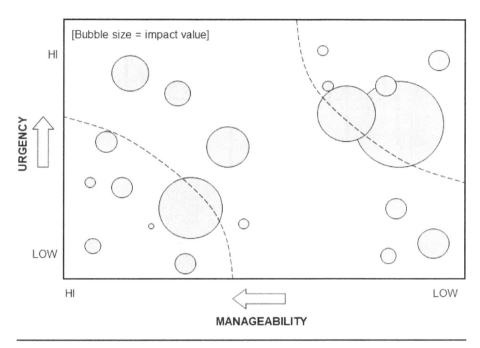

**Figure 6-4** Example Bubble Chart (Urgency, Manageability, Impact)

easy to manage. The top-priority risks are large bubbles (big impact) in the top right-hand corner (very urgent and hard to manage), and we need to give these our immediate attention in order to work out how to address them before they happen. If a large bubble appears in the lower left-hand area, it is lower priority because we have plenty of time to work out what to do (low urgency) and the risk is easy to address (high manageability), even though it would have a major impact if it were to happen. We may wish to prioritise medium-size bubbles with high urgency if they can be managed easily (quick wins), and so on. This type of analysis produces the following risk ranking:

- Top priority:     High impact, High urgency, Low manageability
- Second priority:  High impact, Low urgency, High manageability
- Third priority:   Medium impact, High urgency, High manageability
- . . .
- Lowest priority:  Low impact, Low urgency, High manageability

In order to use a bubble chart effectively, we need to define the ranking order clearly before embarking on the analysis.

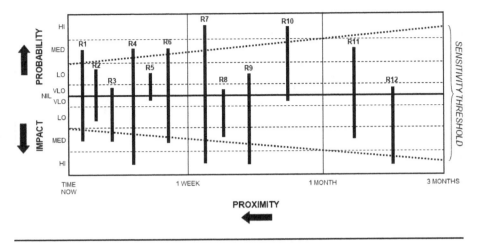

**Figure 6-5** Example Risk Prioritisation Chart (Probability, Impact, Proximity)

The risk prioritisation chart (first devised by Richard Barber in 2003*) uses the upper and lower sections of the *y*-axis for two parameters, with a third along the *x*-axis. Each risk is plotted as a vertical bar. The chart has a "sensitivity threshold", so that risks lying outside the threshold are prioritised for attention and action. In the example in Figure 6-5, probability and impact are shown on the *y*-axis, with proximity along the *x*-axis. As proximity reduces and the potential occurrence of a risk gets further away in time, the sensitivity threshold is set higher, so that only risks with high probability and impact are prioritised if they are far away. The tighter threshold at high proximity means that smaller risks require attention and action when they are closer in time. Other sets of three criteria can also be used in the risk prioritisation chart—for example, risk score, strategic impact, and connectedness.

## TYPICAL TECHNIQUES FOR CATEGORISING THREATS

It's common to limit analysis in the qualitative risk assessment stage of the risk process to simply prioritising individual risks. However, we can gain additional useful information by considering whether there are any patterns among the identified risks. Are there *common causes* that give rise to several risks? Are there *hotspots of risk exposure* in our project that could be affected by more than one risk? Are there *particular phases or periods* during the project that are more exposed than others?

---

* (Barber, 2003)

This information will help us to design effective generic risk responses that tackle groups of related risks, rather than treating each risk on its own.

## Using hierarchical project breakdown structures

Each risk has many different characteristics, several of which can be used to group related risks together. The easiest way to categorise risks is to map them into the various breakdown structures that exist in our project.* These provide natural frameworks for discovering where risks share common characteristics, and they have the advantage of being project-specific, helping us to explore the particular set of identified risks on our unique project. Examples include the following:

- Risk Breakdown Structure (RBS)—potential sources of risk (see Figure 4-1 for an example RBS)
- Work Breakdown Structure (WBS)—areas of project work
- Cost Breakdown Structure (CBS)—breaking the project budget into specific line items
- Organisational Breakdown Structure (OBS)—responsibility for elements of project work
- Product Breakdown Structure (PBS)—project outputs in terms of deliverables
- Benefits Breakdown Structure (BBS)—project outcomes in terms of benefits to stakeholders

Other project hierarchies may exist, and the set of categories to be used for our project will have been defined in the RMP as part of tailoring the risk process (see Chapter 4).

The categorisation process is the same, whichever breakdown structure is used as a framework. We consider each identified risk in turn and decide where it fits in the framework. Having placed all the risks, we then examine the distribution to look for areas where risks are concentrated. We can do this in three ways, depending on the level of qualitative risk assessment that we've conducted.

- We could just count the number of risks in each element of the framework, and see which elements has the highest number. This is simple to do, but it doesn't take account of the importance of each risk. For example, we might find that work package WP1 in the WBS has 10 risks linked to it,

---

* (Hillson, 2007b)

and WP2 has only three. The first impression is that WP1 has a higher risk exposure than WP2. But if all the 10 risks that could affect WP1 are low-priority, and all three risks associated with WP2 are high-priority, the converse could be true.
- We can overcome this weakness by introducing a simple scoring scheme, perhaps saying that each Red risk scores 3 points, a Yellow risk is worth 2 points, and just 1 point for a Green risk. Then we determine which framework element has the highest total score.
- A similar weighting approach is possible if we've used the Probability-Impact (P-I) risk scoring system described in the previous section (see Table 6-3 and Figure 6-3), or something similar. Instead of just scoring each risk as 1, 2, or 3, depending on its prioritisation zone, we can use the unique risk score associated with the combination of probability and impact for the risk (Figure 6-3). We calculate the total score of each framework element in the same way, but the result is more refined and precise.

Mapping risks to each breakdown structure provides useful information on the way identified risks might combine to affect our project:

- RBS—this allows root-cause-analysis, revealing common causes of risk. For example, we may discover that 30% of our risks relate to resource issues, or 15% arise from aspects of the technical solution, while there are apparently no risks from procurement.
- WBS—this indicates areas of the project that are most exposed to risk, and which would bear the greatest impact if risks occurred. We may decide to concentrate our best resources on these areas, or include additional float and contingency, or adopt a different development approach.
- CBS—identifying which line items are most risky will help us to allocate contingency funds or take other measures to protect the budget.
- OBS—understanding who is responsible for risky areas of the project will help us to assign risk owners.
- PBS—knowing where variability in project deliverables could occur will guide us in design, development, testing, and audit activities.
- BBS—identifying possible deviations in project outcomes will help us to manage stakeholder expectations and take preventive action to avoid failure to deliver value.

It is also possible to combine two breakdown structures to allow a deeper analysis of risk groupings across the project. One example is the Risk Breakdown Matrix (RBM),[*] which combines the RBS with the WBS. This allows a

---

[*] (Hillson, Rafele, & Grimaldi, 2006; Rafele, Hillson, & Grimaldi, 2005)

Picking Winners 119

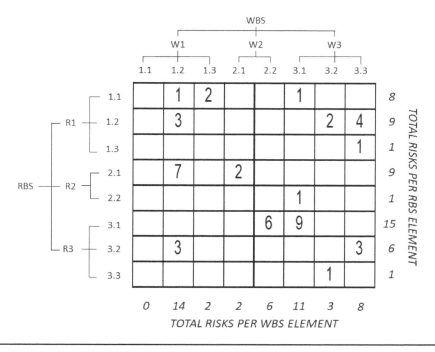

**Figure 6-6** Example Risk Breakdown Matrix (RBS/WBS)

cross-analysis of where each risk comes from (RBS element) and which part of the project it affects (WBS element). We can then see if there are any concentrations in which a particular source of risk touches a specific area of the project. An example RBM is shown in Figure 6-6. This shows that Work Package 1.2 (WP 1.2) could be affected by the highest number of risks (14 risks), followed by WP 3.1 (11 risks). We can also see that RBS element 3.1 gives rise to more risks in this project than any other source (15 risks). Finally, we discover that the biggest single concentration of risks in this project is where RBS element 3.1 affects WP 3.1 (9 risks). We should investigate this further to find out why, and see whether we can design specific generic risk responses to tackle this combination of risk source and area affected.

## Timing issues

Assessing probability and impact tells us how big each risk is, and categorising risks tells us about concentrations of risk that could affect our project. We may also consider other characteristics of identified risks, such as the degree of strategic impact, or how connected each risk is with other risks, or how much each risk matters to stakeholders. This information helps us to *pay attention* to the most important risks, as shown in the top part of Figure 6-2.

But there is another reason for prioritising risks: to tell us where we should be *taking action* to address risks. Figure 6-2 also lists some characteristics of risks that relate to this goal, including proximity and urgency. It's important for us to consider these timing issues as we decide which risks to address first.

- *Proximity* describes how soon this risk is expected to occur, if it does occur, and this is known as the "impact window". If a risk might happen soon (high proximity), then we need to prioritise it for action, if possible.
- *Urgency* describes how soon responses must be implemented in order to address this risk effectively, called the "action window". If we need to act quickly (high urgency), then this is a high-priority risk.

In addition to considering these timing issues in isolation, we must also examine the relationship between proximity and urgency for each risk. A risk with high proximity and high urgency should clearly have the highest priority when we are planning action—it is due to happen soon, and we have to act quickly if we want to manage it. But we also get interesting and useful information by looking at whether impact window and action window overlap or are separated by some distance in time.

Figure 6-7 shows an example in which action window and impact window for each risk are overlaid on a time-sequenced chart. This illustrates a different picture for each of the six risks shown:

- The action window for Risk 1 starts now, and the action must be in place within three weeks. We then have a further five weeks before the risk might occur, giving us some slack. We're also quite uncertain about the precise timing of and potential impact from this risk—the impact window suggests that it might occur anywhere within a 20-week timeframe.
- Risk 2 has a longer action window within which we can implement our chosen actions, although we can't start for two weeks. But there is no gap between the action window and the impact window, so our actions must be completed by the end of the action window.
- For Risk 3, there is only one week during which we are able to influence this risk, even though the risk is not expected to occur for some time. If we miss this chance to address the risk, we may have to rely on good luck!
- The action window for Risk 4 overlaps the impact window, so although in theory we have up to seven weeks to take action, we actually have to act within the first four weeks of the window.
- The period during which action for Risk 5 might be taken starts three weeks after the impact window for the risk opens. This obviously gives

us a problem—the risk might happen before we have a chance to do anything about it.
- Risk 6 has a long action window, and therefore there should be no reason why the action cannot be successfully implemented before the risk might occur.

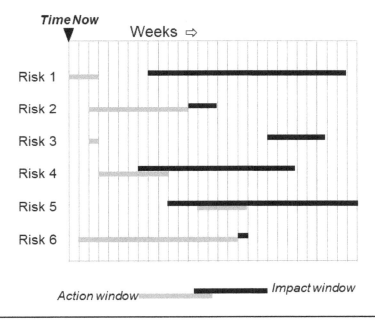

**Figure 6-7**   Overlapping Action Windows and Impact Windows

## MODIFYING THREAT TECHNIQUES TO PRIORITISE OPPORTUNITIES

We've described the approaches for prioritising threats in some detail, showing how to pick out the most important threats where we need to pay *attention*, as well as determining the ones where we need to take focused *action*. We now turn to consider how to prioritise opportunities, and this is where we benefit from one of the important risk concepts discussed in Chapter 1:

*A threat is the same as an opportunity; the only difference is the sign of the impact.*

This means that all of the techniques described above for prioritising threats can be used for opportunities, with only minor modifications. Of course, the prioritisation focus is different: Instead of looking for the *worst threats*, we are seeking the *best opportunities*. But this difference arises entirely from the impact dimension, which is negative for threats and positive for opportunities. Every other characteristic listed in Table 6-2 is the same for threats and opportunities. Opportunities have a probability or frequency of occurrence, and we can assess their degree of connectedness, propinquity, proximity, urgency, manageability, and controllability. If we assess these factors, we can then use the same set of approaches to prioritise and categorise opportunities for our *attention*, as well as looking at the timing issues that might drive our *action*. We want to focus on the best opportunities, as well as developing effective responses to give us the best chance of capturing them.

Consequently, when we want to prioritise opportunities, we can use a P-I Matrix (with or without risk scoring), bubble charts, risk prioritisation charts, mapping into various project hierarchies (RBS, WBS, CBS, etc.), or overlapping impact/action windows, as described above. All these techniques are useful for prioritising opportunities in exactly the same way as they are used for threats. The one exception where we might want to modify the threat-based prioritisation technique is the P-I Matrix.

## Adapting the P-I Matrix for opportunities

Because the main difference between threats and opportunities is the impact dimension, either negative or positive, the main prioritisation technique that needs to be modified for opportunities is the P-I Matrix, where impact is obviously a major consideration.

When we are considering threats only, the P-I Matrix is a simple grid with probability/frequency of occurrence in one direction (usually the *y*-axis), and severity of negative impact in the other direction (*x*-axis). Each threat is plotted on the grid in the appropriate cell and placed into a prioritisation zone, usually Red (high-priority), Yellow (medium-priority), or Green (low-priority). An example is shown in Figure 6-1, above.

We can use exactly the same approach to prioritise opportunities on a P-I Matrix, by simply relabelling the *x*-axis to represent positive impact. Then opportunities in the Red zone are top priority because they are easy to obtain (high probability or frequency means they are likely to happen), and they would have a highly positive impact on achievement of project objectives if they did occur. These "golden opportunities" should receive our attention and action so that we can convert them to real savings and benefits for our project. Opportunities in the Green zone are lowest priority because they are unlikely to happen, and if they did occur then their positive effect is only small. Green opportunities therefore don't require the same level of attention or action. Yellow-zone

Picking Winners 123

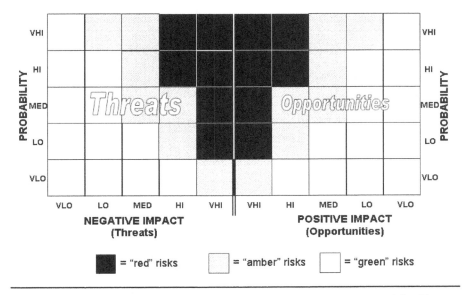

**Figure 6-8** Combined "Mirror" Probability-Impact Matrix (P-I Matrix) for Threats and Opportunities

(medium-priority) opportunities should be monitored to detect whether they become higher-priority and so demand more focused attention.

Similarly, we might choose to use a risk scoring system for opportunities that matches the one used for threats, to allow more detailed prioritisation within the zones, as shown in Table 6-3 and Figure 6-3.

If we follow this approach, we would have two identical P-I Matrices, one for threats and another for opportunities, with the only difference being the sign of the impact on the $x$-axis. As discussed in Chapter 4, however, it is more common to combine the two matrices into a single grid, with the opportunity side flipped over to create a symmetrical "mirror matrix". This is shown in Figure 6-8 (reproducing Figure 4-5). The main benefit of this approach is to bring together the two Red zones into the centre, placing the worst threats and the best opportunities in a single central focus area. These risks together form the most important risks that demand our attention.

The format of this mirror matrix is defined in the RMP for our project (see Chapter 4), including the placement of thresholds between Red/Yellow/Green zones.

## WRITE IT DOWN

After we've completed this step in the risk process, we'll have lots of extra data on the risks we've identified. We've assessed how likely each risk is to occur

Table 6-4 Risk Data to Record after Qualitative Risk Assessment

| Data Field | Minimum | Typical | Detailed |
|---|---|---|---|
| **Qualitative Risk Assessment Data** | | | |
| Probability/frequency of occurrence (current, pre-response) Qualitative (for example High, Medium, Low) Quantitative (for example % range, frequency) | X | X X | X X |
| Impact on each project objective (current, pre-response) Qualitative (for example High, Medium, Low) Quantitative (for example, three-point-estimates of days, dollars, technical performance measures) | X | X | X X |
| Overall risk ranking Red/Yellow/Green (or similar) Risk score (calculated from probability and impact) Risk score (using more complex algorithm) | X | X X | X X X |
| Other prioritisation parameters (urgency, manageability, detectability, proximity etc.) | | | X |
| Risk source (RBS element) | | X | X |
| Project area affected (WBS element, CBS element etc.) | | | X |
| Impact window | | | X |
| Action window | | | X |
| Related risks | | | X |

(probability/frequency) and its potential impact on our ability to achieve our project objectives (negative for threats, positive for opportunities). We may have considered other important risk characteristics, including some of those in Table 6-2. Perhaps we've assigned our risks to categories in an RBS or mapped them into other project hierarchies, and we may also have considered various timing issues such as the impact window or action window for each risk.

All these data need to be recorded in a common format so that we can use them in the later stages of the risk process. After the risk identification step (see Chapter 5), we initiated the risk register to hold data on all identified risks. The level of detail in a risk register can vary, and this is defined for our project in our RMP. The data fields to be completed in the risk register following qualitative risk assessment are shown in Table 6-4, for a minimalist, typical, or detailed risk register format.

These new data will allow us to generate useful information about the riskiness of our project, which will shape the next steps of the risk process. For example, we can discover the worst threats and the best opportunities, based on the mapping of each risk in the P-I Matrix. At its simplest, we pick risks in the top Red prioritisation zone, or we might have a more detailed ranking if we've used risk scoring. From these data, we can generate a "top risk" list of the risks that demand the highest level of attention. Two issues arise here:

- Firstly, how many risks belong in a "top risk" list? Many organisations ask for a "Top 10", but what if the risk ranking eleventh is a showstopper? Or we might take the "Top 20" and perhaps include lots of risks that aren't too important. It's better to define a threshold for inclusion in the "top risk" list—for example, all Red risks, or all risks scoring over a set figure. If this is only three risks or as many as thirteen, so be it; these are the risks that demand our attention and action.
- Secondly, should we have two lists: one giving "top threats" and another for "top opportunities", or should there be one overall "top risk" list? To answer this question, we need to remember the purpose of prioritisation. Risk identification usually identifies more risks than we can manage within our limited time, effort, and resources, so we need to determine which risks are more important. This will inform the remainder of the risk process, particularly when we come to plan appropriate risk responses. If we list top threats and top opportunities separately, we can focus attention and action on the worst threats and the best opportunities. But what if all the highest-ranking risks are opportunities? By focusing on the top few risks from each list, the top risks from one list might obscure the middle ones from the other list. It's better to create a single list of "top risks", interleaving threats and opportunities as necessary, then focusing on the top ones regardless of their type. If this reveals that all the most important risks are opportunities (or all threats), then so be it; these are the risks that we must focus on.

## SUMMARY AND REFLECTION QUESTIONS

If we succeed in identifying the real threats and opportunities facing our project, our next task is to prioritise them for *attention* and *action*, deciding which are the most important and which require addressing first. Usually people rank risks using just probability/frequency and impact, focusing on the ones with the highest likelihood of occurring and the highest potential effect on achievement of objectives. But there are several other reasons that we might regard a risk as important, not just probability/frequency and impact. These could include the level of impact that a project risk might have on strategic goals or whether a risk is highly correlated with other risks.

Whichever factors we use to rank risks to decide which demand our *attention,* we can use the same approaches for both threats and opportunities, because they are essentially the same, apart from the type of impact (negative for threats, positive for opportunities). The most important risks are usually defined as the worst threats and the best opportunities, as ranked by probability/frequency and impact. Prioritisation techniques using other risk characteristics also work for both types of risk.

Similarly, we can group risks together in various ways to expose common causes of risk or hotspots of risk exposure, and we can do this in the same way for both threats and opportunities.

But when we want to discover the highest-priority risks in terms of taking *action*, these are not always the biggest ones. In addition to size, we also need to know how soon the risk might occur and how quickly we need to act if we are to respond effectively. Or we might consider how easy it is to influence whether a risk will happen, or whether we can control the size of its impact. We need to prioritise risks for both attention and action, and this requires consideration of different factors.

### Reflection questions

- What is the difference between probability and frequency, and when do you use each one?
- List and define the key characteristics of a risk that might be used for prioritisation.
- How would you define the "best opportunity" on your project? Is "best" defined only by its probability/frequency and impact, or are there other factors to consider?
- Why is prioritising risks for attention not the same as prioritising them for action? Give examples of where these might be different.
- *"A threat is the same as an opportunity; the only difference is the sign of the impact"*. Discuss why this is helpful when we are prioritising risks for further attention and action.

## THE NEXT STEP ("NOW WE'VE FOUND THE IMPORTANT OPPORTUNITIES, WHAT CAN WE DO ABOUT THEM?")

At the end of the qualitative risk assessment step, we have a good idea of which risks are most important, both in terms of demanding our attention and requiring our action. This includes the highest-priority threats as well as the top opportunities.

For some projects, especially those that are complex, innovative, or strategically important, deeper analysis of the project's exposure to risk is required. In these cases, the next step in the risk process is quantitative risk analysis, using mathematical simulation models to explore the combined effect of identified threats and opportunities on overall project outcomes. Quantitative risk analysis is described in Chapter 7. When this in-depth analysis is complete, we can then move on to risk response planning (Chapter 8).

For other projects, we may get enough information from qualitative risk assessment to move directly to deciding what to do about the risks we've identified and assessed. In these cases, deeper quantitative analysis is not needed, and we are ready to start actually managing risks, instead of just identifying and assessing them. So far in the risk process, through risk identification and qualitative risk assessment, we've been gathering useful information to ensure that we properly understand the real risks to our project. But mere understanding doesn't change anything if we don't do something different in response to the new information we've obtained. It's time to plan how to respond to our threats and opportunities!

# Chapter 7
# Using Numbers to Model Opportunities

Our exploration of the practical side of opportunity management has so far not used any statistics. After the initial risk management planning stage, we discovered how to find opportunities and describe them clearly, and how to pick winners with a range of prioritisation techniques. On some projects, we will now have enough information to start planning and implementing responses to these opportunities, alongside the threats. But the risk challenge on other projects may demand a more in-depth analysis of the combined effect of risks on project outcomes. These projects make use of quantitative risk analysis, and if we are including opportunities in our risk process, then we need to know how to **use numbers** to model them. This chapter explains how.

## PURPOSE AND PRINCIPLES OF QUANTITATIVE RISK ANALYSIS

The main output from the preceding step in the risk process—qualitative risk assessment—is a prioritised list of threats and opportunities. These are usually ranked by probability/frequency and impact, although other factors can be used in addition to determine which are the most important risks that demand our attention and action. As well as the prioritised list, it is common to group risks together using project hierarchies—Risk Breakdown System (RBS), Work Breakdown System (WBS), Cost Breakdown Structure (CBS) etc.—in order to reveal common causes of risk or concentrations of risk exposure in various elements of the project.

Qualitative risk assessment forms the backbone of the risk process for many organisations, because it focuses further attention and action on the risks that matter most. However, it suffers from one major shortcoming: *qualitative risk assessment only deals with individual threats and opportunities.* It is impossible to work out the *overall riskiness of the project* by looking at the set of individual risks that have been identified and assessed. There is no reliable way of determining what effect these risks might have on the final project outcome. For example, risks with schedule impact may not affect the overall project duration if their impact would occur off the critical path. Individual risks are also not independent: the occurrence of one risk might make another more likely, or less severe, or impossible.

Even grouping individual risks together using different categories doesn't tell you how risky the project is as a whole. The number of possible combinations of identified individual risks is just too large for us to deal with. For example, if our project has just two risks (Risk A and Risk B), there are four possible combinations of these risks that might affect the project outcome: no risks occur, Risk A occurs, Risk B occurs, both Risks A and B occur. With three risks (A, B, C), there are eight possible combinations (none, A, B, C, AB, AC, BC, ABC). With 20 risks there are over a million possible combinations! Even if we remove infeasible combinations of risks, or take account of dependence between risks, the number of possible outcomes for the typical project is far too high for us to understand.

And yet we need to answer the question, "How risky is this project?"[*] This requires a different set of techniques, which fall under the heading of *quantitative risk analysis* (QRA).[†] The riskiness of our project arises from the set of individual threats and opportunities to which the project is exposed, but there are other forms of uncertainty that matter when we are considering the project as a whole. QRA needs to address all forms of uncertainty, not just individual risks, in order to *provide a reliable evaluation of the overall risk exposure of the project,* and to examine how these various types of uncertainty might interact within the project to affect the final outcome. ***The purpose of QRA is to evaluate overall project risk by considering the combined effect of uncertainty on project outcomes.***

There are a number of important ***principles*** to remember when using QRA.

## QRA is not always needed

Not all projects require the use of QRA, and the requirement will be defined for each project in its Risk Management Plan (RMP). If a project is small or simple

---

[*] Hillson, 2014b
[†] (Vose, 2008)

or low-risk, then perhaps sufficient insight to support effective management of risk can be gained from qualitative risk assessment alone. But *large, complex, high-risk, or strategically important projects usually benefit from the application of QRA techniques.*

The problem here is one of definition: When is a project "large, complex, high-risk, or strategically important"? Each of these parameters depends on the organisation type and the industry sector, as well as the inherent characteristics of the project itself. The way that each parameter is defined will vary from one organisation to another, and maybe also between different project types within the same organisation. It is clearly important that the project sponsor and those authorising the project understand these scaling factors, and that they can state unambiguously where a particular project sits in the spectrum of size, complexity, riskiness, and strategic importance. They will then be in a position to determine whether the use of QRA is appropriate for this project.

It's also possible that some small, simple, low-risk, tactical projects may decide to use QRA, but this should be done for a specific reason and not just to comply with a one-size-fits-all risk approach.

Whatever the project size or complexity, the easiest way to determine overall project risk exposure is by using QRA, and any project might decide to use it for this reason alone.

## Where QRA is required, know why and how to use it

If the RMP indicates that this project should be using QRA, there are some important questions to answer before we start on any analysis. Why do we need to do this analysis? What questions are we trying to answer? The questions to be answered should be clearly defined at the start of the QRA step. For example, do we want to know how risky the project is as a whole so that we can make a "go/no-go" decision, or work out how much contingency we need? Or are we more interested in assessing what range of outcomes are possible, or trying to find the biggest risks? Understanding the specific reason for QRA will allow the project to set clear objectives and scope for its implementation, and it will also help us to define what outputs are required to answer the questions.

Once we know the purpose of conducting QRA on our project, we can design a risk model that addresses this purpose. We need to be aware of the questions that we're trying to answer, then build the model specifically to provide the data to answer those questions.

The risk model might be limited to examining the uncertainty associated with a single key project variable such as schedule, cost, resource, margin, profitability, cashflow, internal rate of return (IRR), net present value (NPV), etc. Alternatively, the QRA risk model can support an integrated multivariable analysis, perhaps addressing both schedule and cost together while reflecting resource constraints. The scope of the risk model should also ensure an appropriate level of detail—for example, addressing whole-project outcomes, or focusing instead on one key project phase or a particular subproject.

Finally, if we're using QRA, we must also make sure that it is properly resourced, with skilled people, appropriate tools, and available time and effort for data gathering, data validation, and the analysis and interpretation of results. A wide range of proprietary risk tools is available, or a risk model can be created in common office software, and we should use a tool that matches the level of analysis we are doing.

## No models are "correct"

The British statistician George Box famously reminds us that *"All models are wrong; some models are useful"*, and that, *"The most that can be expected from any model is that it can supply a useful approximation to reality."*\* Of course, when we are building risk models, we aim to be as realistic as possible, without getting into too much detail. But Box's warning is particularly applicable when we're modelling the effect of uncertainty—we use QRA precisely because we don't know every detail about our project with perfect certainty.

QRA is based on a model of the project into which uncertainty is added. This usually starts from an existing baseline, such as a project plan or budget, which can be quite detailed, or we might use a summarised version. Einstein reputedly said *"Make things as simple as possible, but not simpler"* (although what he actually said was, *"It can scarcely be denied that the supreme goal of all theory is to make the irreducible basic elements as simple and as few as possible without having to surrender the adequate representation of a single datum of experience"*†), and this is the key to a good risk model. The model needs to be a faithful representation of the realities of our project, including task dependencies, key constraints, external inputs, etc. But it also needs to be constructed at a level of detail that both allows the main risks to be shown and also makes their effects visible.

## Include all types of uncertainty

When building the risk model, we must obviously reflect the effect on our project of all identified individual risks, including both threats and opportunities. But we shouldn't limit input only to the top-priority risks, or red risks, or risks

---

\* (Box & Draper, 1987; Box, Hunter, & Hunter, 2005)
† (Einstein, 1934)

with big impacts. All risks should be modelled, because the effect of smaller risks might be cumulative, synergistic, or antagonistic.

Individual risks are not the only source of uncertainty that could affect project outcomes. We must also model other types of uncertainty, including *variability* in planned activities and tasks (using ranges of values), as well as *ambiguity* about future decisions and options (using stochastic branches). We also need to identify *dependencies between risks* (using correlation). Data are usually based on the current risk register, which provides an important audit trail.

Finally, we must remember that uncertainties are often connected or related, and not independent. For example, the occurrence or non-occurrence of some risks might affect the significance of others (either positively or negatively). We need to include appropriate *correlations* in our risk model to account for the way that various uncertainties are related together.

## Use best available data

The quality of the analysis is driven by the quality of the input data. People often hesitate to start QRA if they feel that the data are not of sufficiently high quality. But we shouldn't wait for complete or perfect data—it's never available! The point of QRA is to model uncertainty, so incomplete or unsure data is fine. The well-known acronym GIGO (Garbage In—Garbage Out) puts a lot of people off, because we don't want to get rubbish results by feeding in rubbish data. But we also need to beware of NINO: Nothing In—Nothing Out. If we genuinely don't know something, then we should say so. If our estimates of the possible durations or effort estimates of a task really are wide or woolly, we need to include this in the risk model so that the true effect of our uncertainty can be taken into account.

## Use the results

If we spend time and effort in conducting QRA on our project, we need to ensure that we obtain a good return for our investment. QRA produces lots of valuable information to help us manage the project more successfully. Its main output is an evaluation of *overall project risk exposure* to tell us how risky the project is as a whole. QRA outputs indicate the *likelihood of achieving key project objectives,* and show us the *possible variation in project outcomes* (from realistic minimum to credible maximum, as well as most-likely values). We can discover *which individual risks have the most influence* in driving overall project cost or duration, *which are the most uncertain parts of the project,* and which risks and planned activities *influence the critical path* of the project.

Despite the availability of all these useful outputs, there's no point in doing QRA if we don't use the results to influence project strategy and tactics. We need risk-based decision-making at all levels of the project, from the big decisions about development options or project prioritisation, down to routine decisions on resource allocation, change management, or team motivation.

We can also use QRA results to inform risk response development, enabling us to focus on the main drivers of outcome uncertainty, as well as the key individual risks. As we consider alternative risk response strategies, we can use the model to test their relative effectiveness in changing overall project risk exposure by performing before-and-after analyses.

One final health warning is important when we come to use the results of QRA. We must always take the results as they are and not try to force the model to produce the "Right Answer"! The model outputs will reflect the input data, especially the uncertainty associated with identified risks as well as sources of variability and ambiguity. We may initially be surprised by the results, but that doesn't mean they are wrong. We should of course be careful to understand how the risk model works and why certain results have been reached, and we must validate its functionality. But we shouldn't assume that the model is producing "wrong results"—perhaps our project really is more (or less) risky than we thought! Risk models deliver ranges of possible outcomes, based on the uncertainties that have been modelled, and interpretation of the results is still needed. We must be careful not to overinterpret—if the inputs are uncertain then the outputs will necessarily be uncertain too. The results cannot be more certain than the data!

## Summary

The purpose and principles of quantitative risk analysis are summarised in Table 7-1.

Table 7-1 Purpose and Principles of Quantitative Risk Analysis (QRA)

| Purpose | To evaluate overall project risk by considering the combined effect of uncertainty on project outcomes |
|---|---|
| Principles | • QRA is not always needed<br>• Where QRA is required, know why and how to use it<br>• No models are "correct"<br>• Include all types of uncertainty<br>• Use best available data<br>• Use the results |

## TYPICAL TECHNIQUES FOR MODELLING THREATS

The most frequently used QRA technique is Monte Carlo simulation, which is covered in this chapter, although other approaches—decision trees, influence diagrams, system dynamics modelling, multi-criteria decision analysis, real options analysis, etc.—are useful in some circumstances.

### Monte Carlo simulation

Monte Carlo simulation is the most common way to analyse risk using numbers. But many people view QRA as too difficult, perhaps because it involves mathematics, statistics, and computers. As a result, they miss out on the insights available from this powerful technique.

Readers needing detailed guidance on how to use Monte Carlo simulation for QRA can find help in the many textbooks and training courses that cover this specialised topic,[*] and we won't repeat the detail here. In summary, a number of key steps are involved in conducting a Monte Carlo QRA effectively:

- *Define the purpose of the analysis.* The particular emphasis and purpose of quantitative risk analysis for this project will be defined in the project's RMP. The scope of the analysis might only cover schedule risk or cost risk, or an integrated view may be needed. Quantitative risk analysis can also be applied to other objectives, such as IRR or NPV.
- *Develop the risk model.* For a quantitative schedule risk analysis,[†] the risk model is usually based on the project schedule (i.e., the critical path network), either in its full form or summarised (depending on the number of activities). The most realistic results are obtained if the baseline schedule is resource-constrained, and pinned dates must be removed when the analysis is run in order to allow the full effect of duration uncertainty to be exposed. A quantitative cost risk analysis model is usually based on the detailed project estimate or CBS, which is often set out in a spreadsheet. A single integrated risk model can be created for analysis of both schedule and cost risk by using the critical path network and ensuring that all project costs are included within the schedule, preferably along with resource information.[‡]

---

[*] (Hulett, 2009, 2011; Vose 2008)
[†] (Hulett, 2009)
[‡] (Hulett, 2011)

- *Generate input data and enter into risk model.* Once the initial baseline risk model has been developed, the data required for the analysis can be derived and input. These data must represent *identified threats and opportunities,* as well as other types of uncertainty, including both *variability* (presented as ranges of values for planned activities) as well as *ambiguity* (the possibility of alternative options, modelled using stochastic branches). Interactions between sources of uncertainty must also be included, using statistical *correlation* between the ones that are linked.
- *Initial analysis—Validate risk model and produce results.* The risk model is validated by running a limited number of iterations, checking that no structural errors exist, no data input errors were made, and nothing illogical has been included. Any errors should be corrected before proceeding further. The validated risk model is then run through a full analysis with a large number of iterations, allowing the Monte Carlo simulator to sample randomly from the input data and calculate the range of possible outcomes that reflect the sources of uncertainty in the model.
- *Secondary analysis—Run risk model including risk responses.* The risk model can be modified to include the effect of risk responses and actions. Comparing the results from this secondary analysis with the initial results allows us to demonstrate the effectiveness of planned responses, and we can determine whether additional risk responses are needed.
- *Produce and interpret analytical outputs.* The final outputs from the analysis will present the range of possible outcomes, allowing an evaluation of overall project risk exposure, as well as providing an assessment of the likelihood of achieving project objectives and exposing the main drivers of uncertainty in project outcomes.
- *Decide appropriate course of action and report results.* The outputs produced should be carefully considered and the need for any resulting actions decided upon. Actions could include anything from a complete re-strategising of the project to minor tweaks in the logical sequence of the project's activities. As a final step in the process, a report should be produced that details the analysis that has been undertaken including the results obtained and the resultant decisions or recommended changes.

We could add much more detail to describe each of these steps, but our purpose here is to understand how to use QRA to reflect the combined effect of all sources of uncertainty on our project, including upsides as well as downsides. Consequently, we'll focus on how to generate input data that reflect uncertainty, and how to interpret analytical outputs. In this part of the chapter, we describe how typical threat-focused QRA is done, then in the following part we'll widen the discussion to cover opportunities.

## Generating input data

The input data used for our QRA risk model needs to account for all types of uncertainty that might affect the outcome of our project. This includes:

- The possibility of variation in estimated durations, costs, and resource requirements for the planned activities in our project (*variability*)
- Future areas of the project where we currently lack clarity on how to proceed (*ambiguity*)
- The *individual project risks* that we've identified during the risk process

Each of these three types of uncertainty are handled differently when we produce input data for the risk model. We also need our risk model to reflect the reality that different sources of uncertainty can interact, either by acting in synergy to make their combined effect bigger, or by countering each other to make the overall effect smaller. These interactions are modelled using *correlation*.

- *Variability*. When we create the project plan, we develop the WBS to its lowest level, describing the elements of work to be performed, then we estimate various parameters characterising each lowest-level activity, including durations, costs, and resource requirements. Although we all know that actual performance is extremely unlikely to be exactly the same as our estimates, it is common for the project plan to be made up of single-point estimates. This makes the plan appear to be more certain than it is in reality. In fact, it's very likely that any given element of project work might take more or less time than we estimated, and/or cost more or less, and/or need more or fewer resources. So instead of providing single-point estimates, we could reflect this inherent variability using ranges of values.

    The most commonly used way to reflect the possibility of variation in planned values is the *triangular distribution* (also called a *three-point estimate*). This replaces a single-point estimate with three values: minimum, most-likely, and maximum. The most-likely is our best estimate of what will actually happen in reality. It's actually almost impossible to estimate a theoretical minimum or maximum for most activities, so instead we aim to estimate credible or realistic minimum and maximum values, with a small chance of the actual values lying outside the min–max range. Many people try to work with a 5% possibility of reality being less than the estimated minimum and 5% higher than the estimated maximum. This modification of the three-point estimate is called a *modified triangular distribution*.

    Other types of range are possible for which the three-point estimate is not suitable. For example, if we are genuinely unable to estimate a most-likely value, we can use a *uniform distribution,* which sets credible minimum and maximum values and assumes that all interim values are equally likely. If we think that variation in value is random, we might use a *normal distribution.* With more detailed information on how a value might vary, we could perhaps replace the three-point estimate with some kind of curve, such as a *beta distribution* or a *lognormal distribution.* If our activity can only take specific discrete values—for example, a trial which might last one week, two weeks, or three weeks, but never a part-week—we could use a *spike distribution* (also called a *discrete distribution*).

These various types of range are illustrated in Figure 7-1. They are used in the risk model to replace single-point estimates of duration or cost or resource requirement, reflecting variability in planned activities.

- *Ambiguity.* Most projects include aspects for which we're not certain about how things might work out in future. For example, if we're undertaking an acceptance trial, we may pass the trial first time or we may need to repeat it once or twice before we can proceed to delivery, and after three failures we might have to cancel the project. If we need planning permission before we can start construction, our design may be approved without comment, or approval may be subject to amendments, or our application may be rejected and require resubmission. We might decide to outsource part of our project that involves development of novel elements if we can find a suitable specialist partner by the time the work needs to start; otherwise, we might have to do it in-house.

  Each of these examples illustrates a type of uncertainty that can affect the project as a whole, including the overall duration and cost. But they are not the same as individual risks, which either happen or don't. These uncertainties represent alternative futures, different ways in which things might turn out. This type of uncertainty can only be modelled using *stochastic branches.* These are used in a risk model to introduce alternative logic into the project schedule or optional cost elements into the project budget, which are sampled in the risk model with a frequency that matches the probability of occurrence of the risk. Stochastic branches are characterised by uncertainty—sometimes they occur and sometimes they don't. When the Monte Carlo simulation is run, each iteration either includes the stochastic branch or doesn't, depending on the conditions that are set for its occurrence.

  There are two types of stochastic branch: *Probabilistic branches* have an associated likelihood of occurrence, and *conditional branches* are only included if pre-set conditions are met. Each type is useful for modelling ambiguity under different circumstances. Taking the three examples mentioned above, we can estimate the probability that an acceptance trial needs to be run once, twice, or three times, and model that using a

Using Numbers to Model Opportunities   139

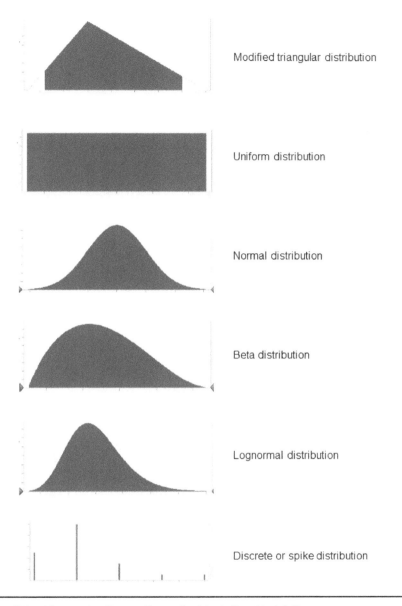

**Figure 7-1**   Alternative Range Types for Modelling Variability

probabilistic branch. The three alternative outcomes from the planning permission process each has a different probability of occurrence, and a probabilistic branch can reflect these. Outsourcing novel elements of the work or doing it in-house depends on whether we can select a partner in time, so we use a time-based conditional branch in the risk model that

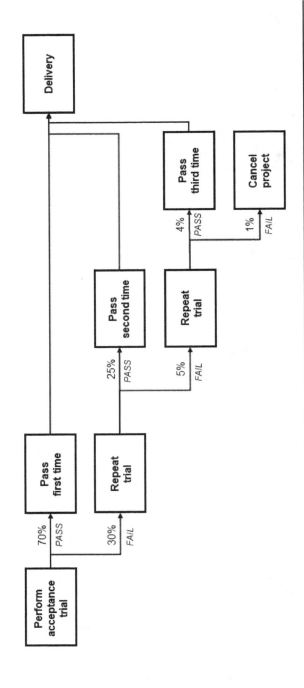

**Figure 7-2** Example Probabilistic Branch—Acceptance Trials

takes the outsourcing route if a partner is found by the deadline, otherwise the in-house route is followed. These three examples are illustrated in Figures 7-2, 7-3, and 7-4, respectively.

- *Individual project risks (threats).* There are two ways to include the effect of individual project risks in a risk model. The one most often used is to map each risk to those project activities or cost elements that would be affected if it occurred, then taking account of the impact when deriving the variability distribution. In this approach, the effect of threats is used to produce the worst-case pessimistic maximum value in a triangular distribution, or in one of the other distribution types described in the previous section. In a similar way, opportunities can be considered when estimating minimum values. While this approach has some merits, it depends on all identified risks being mapped to existing activities in the project schedule or to existing cost elements in the project budget. However, the effect of some risks may lie outside planned activities, and some risks might map to more than one project element, so this mapping approach can get very complicated if it is to be at all realistic.

Fortunately, there is an alternative way to include individual project risks in the risk model, one which does not depend on mapping them to planned activities and including them in the variability distributions. Individual risks can be simply modelled using *probabilistic branches,* in which the probability that determines whether the branch is included in an iteration equals the chance that the risk will occur. If the branch is not triggered, then the iteration does not include the effect of the risk. When the branch is triggered in a particular simulation iteration, the risk impact is included in the risk model as an additional project activity in the schedule or an additional cost element in the budget. This impact can be represented as a single value for the additional time, cost, or resource incurred by the occurrence of the threat, or if the degree of impact is uncertain, then a range estimate can be used, perhaps one of those shown in Figure 7-1. The values of probability and impact used for the probabilistic branch are based on the information held about this particular threat in the project risk register, which was recorded at the end of the preceding process step (qualitative risk assessment).

Use of a probabilistic branch to model a threat with possible impacts on the project schedule is illustrated in Figure 7-5. A probabilistic branch showing the effect of a threat that affects the project budget is shown in Figure 7-6.

Using probabilistic branches to model individual project risks into the risk model is better than including threat impacts into the maximum values of activity estimates, for several reasons:

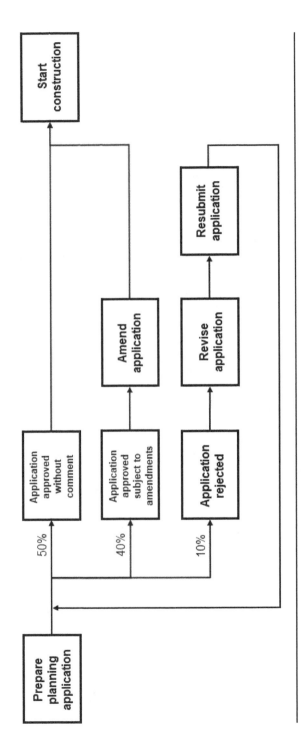

**Figure 7-3** Example Probabilistic Branch—Planning Permission

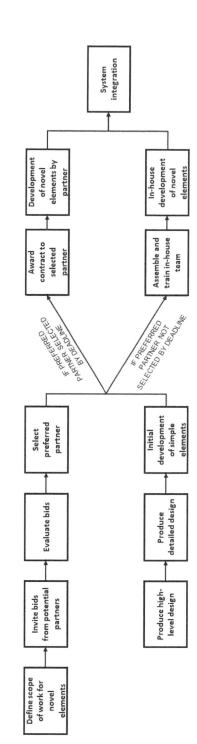

**Figure 7-4** Example Conditional Branch—Outsourced or In-House

144  Capturing Upside Risk: Finding and Managing Opportunities in Projects

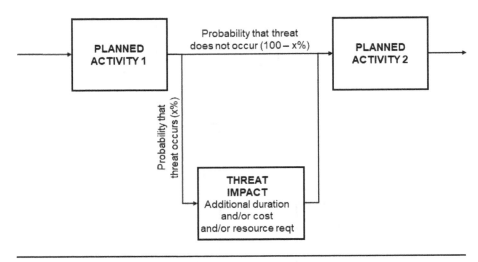

**Figure 7-5**  Probabilistic Branch for Modelling a Threat into the Project Schedule

- o  We can model risks that do not obviously map to existing schedule activities or budget cost items.
- o  We don't have to deal with the situation in which one risk maps to several model elements.
- o  The presence of individual risks in the risk model is clear and explicit, and not hidden within the distributions of other elements.
- o  When we analyse the drivers of uncertainty in overall project outcomes and perform sensitivity analysis, we can see the specific effect of individual risks, indicating which ones have the biggest effect on the project as a whole.
- *Correlation.* Although it may sometimes seem like it to the project team, things don't happen completely by chance on a project. There are limits to

| COST ELEMENT 2.1 | ITEM COST |
|---|---|
| Cost item 2.1.1 | $ A |
| Cost item 2.1.2 | $ B |
| Cost item 2.1.3 | $ C |
| Optional threat cost item:<br>Probability that threat occurs (x%)<br>Probability that threat does not occur (100 – x%) | $ D<br>$ zero |
| TOTAL COST<br>With threat (x% chance of achieving this total)<br>Without threat (100-x% chance of achieving this total) | $ A+B+C+D<br>$ A+B+C |

**Figure 7-6**  Probabilistic Branch for Modelling a Threat into the Project Budget

randomness, particularly later in a project, when the future is constrained by the past. This needs to be reflected in the risk model, otherwise the Monte Carlo simulation will assume by default that all uncertain variables throughout the risk model are unrelated and can vary at random. Risk models use *correlation* between elements to indicate to the simulator that values selected for related elements must be related.

The problem is that the default for all QRA tools is to use no correlation. This means that correlation must be added to the risk model intentionally. Many people don't understand why correlation is vital in producing a realistic risk model, and they don't know how to use it when performing QRA.

Absence of correlation reduces the range of possible project outcomes calculated during the simulation, with random uncertainty in uncorrelated elements cancelling out—good performance or good luck in one area is countered by poor performance or bad luck in another. This leads to unrealistically optimistic results from the QRA.

In reality, risks are interdependent: If one risk occurs, it can make other risks more or less likely, or it can change the potential impact of other risks, or even stop some risks from happening while introducing new risks. Risks can also be related through common causes (if one risk is triggered, it is more likely that others with the same cause will also occur).

Planned activities are also related, with earlier performance setting trends that persist later—for example, if the project team is unmotivated and low-performing in initial phases of the project, they are unlikely to become high-performing later on.

These relationships need to be included in the risk model, creating dependency through *correlation groups* that link activities and risks which can affect each other. In these cases, the ability of the Monte Carlo simulator to sample randomly needs to be constrained. A correlation group identifies elements in the model in which sampled values are related, either positively or negatively, and uses a correlation coefficient (set between –1 and +1, or from –100% to +100%) to model the strength of the relationship. This means that if the simulator selects a particular value in one variable, a similar value will be selected in other related variables, including both risks and planned activities.

The first step in using correlation is to define correlation groups of related risks and activities. Then we can set a correlation coefficient to describe the relationship between members of each correlation group.
- o A correlation coefficient of zero means that there is no connection between group members, and the behaviour of any one member has no influence on any other members.

o Positive correlation coefficients mean that the behaviour of one group member leads to similar behaviour in other group members. Positive correlation between risk probability values means that if one risk occurs, related risks are more likely to occur. Where risk impact values are positively correlated, if the impact of one risk is particularly high, related risks will also have high impacts. The same is true where project activities or cost elements are positively correlated: The value sampled from the distribution will be similar in all cases within a correlation group. A coefficient of 1 indicates perfect correlation between all group members, and lower coefficients have less strong influences.
   o Negative correlation coefficients mean that the behaviour of one group member leads to opposite behaviour in other group members. A high value for a parameter of one group member will lead to a low value for negatively correlated members. A coefficient of −1 indicates perfect matching between all group members, and lower coefficients have weaker influences.

It is rare for members in a correlation group to be perfectly correlated, and there's no reason to set up a group with zero correlation, so in most cases correlation coefficients are set to interim values, commonly either between −0.6 and −0.8, or between +0.6 and +0.8.

## Understanding outputs

Once the risk model is created with data that represent the full spectrum of uncertainties that could affect project outcomes, a Monte Carlo simulation can be performed. This runs through the risk model and randomly samples values from the input data to create one possible result for the project—this is known as an "iteration". The simulation continues to make many iterations, each time taking random samples from the input data, building up a series of possible project outcomes. The combined results from all iterations is then presented using several output formats, including:

- S-curve (cumulative probability density function)
- Sensitivity analysis
- Criticality analysis

These are described below.

- *S-curve.* The main output from a Monte Carlo simulation is a cumulative probability density function known as the S-curve. This may be supported by a histogram which presents the incidence with which each particular result was obtained. The S-curve allows the combined effect of all types of uncertainty to be analysed, including variability, ambiguity, and individual project risks. S-curves can be created for the duration of the overall project, or they can be used to show predicted dates for project completion, interim milestones, sub-projects, or major activities. When analysing project cost, S-curves can reflect the final outturn cost for the whole project or for the cost of major budgetary items or sub-projects. Figure 7-7 provides examples of S-curves for schedule and cost. (Different formats are used to report the results from risk models that integrate both time and cost uncertainty.)

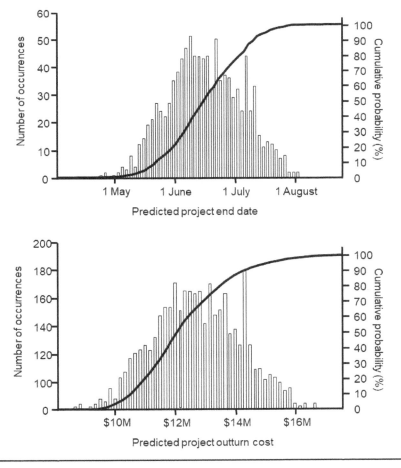

**Figure 7-7**   Example S-Curves for Project End Date and Cost

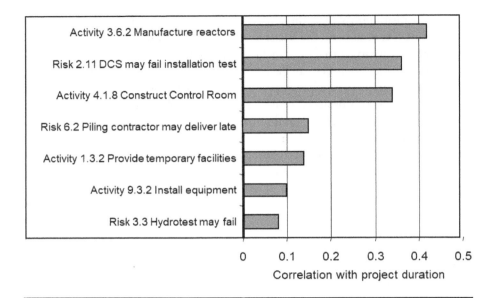

**Figure 7-8** Example Tornado Chart

- *Sensitivity analysis.* This indicates which elements in the risk model have the greatest influence on overall project outcome, by relating the degree of variability in a particular element with the variation in the overall project outcome. Sensitivity analysis can be applied to both schedule and cost-risk analysis and is expressed as a correlation coefficient (from –1 to +1) indicating the relationship between each activity or risk and the total outcome.

  Elements with high sensitivity are key drivers of project uncertainty, because a large change in the element produces a correspondingly large change in the overall project outcome. This applies to risks as well as to planned project activities and cost elements.

  Elements in the risk model can be ranked by sensitivity to indicate which are the most significant causes of uncertainty in the overall result. This information is often presented graphically as a Tornado Chart to highlight the major risk drivers. Schedule activities, cost items, or risks with high sensitivity should be treated with priority when determining areas for further risk management attention and action. An example Tornado Chart is shown in Figure 7-8.

- *Criticality analysis.* Another key QRA output is criticality analysis, which only relates to quantitative analysis of schedule risk.[*] A schedule has at least one critical path, which is the longest route from beginning to end,

---

[*] (Hulett, 2009)

and which determines the overall project duration. During a schedule risk analysis, however, the Monte Carlo simulator makes multiple runs through the project schedule, randomly varying activity durations according to the input data which reflects the uncertainty and mapped risks. Some activities will take longer than the original planned duration, while others will be shorter. As a result, the critical path will almost certainly vary during the simulation, because previously critical activities might be completed in a shorter time while other non-critical activities are extended. In fact, during the many iterations of a risk model, a number of alternative critical paths might be followed.

It is possible to calculate a criticality index for each activity in the risk model, defined as the number of times that activity appears on the critical path, usually expressed as a percentage of the total number of iterations. Therefore, an activity which is always critical has a criticality index of 100%, whereas one which can never be on the critical path has zero criticality. The activities of interest are those with criticality between 1% and 99%, which might become critical under certain circumstances. Ranking activities by criticality index highlights those activities which are most likely to drive the overall duration and completion date, and which therefore require focused risk management attention.

## Using QRA results to evaluate overall project risk

Overall project risk is defined as, "the effect of uncertainty on the project as a whole".[*] If we want to evaluate the riskiness of the whole project, we have to answer two questions, both of which have quantitative answers:

- "How likely is this project to succeed (or fail)?"
- "What is the potential range of variation in outcome?"

QRA allows us to answer these questions because it models the effect of uncertainty on the project as a whole. The main output from QRA is the S-curve, presenting the range of possible outcomes against the probability of each value being achieved, and this contains all the information needed to evaluate overall project risk exposure. The best way to illustrate this is to use a worked example. Figure 7-9 shows an S-curve resulting from a cost-risk analysis, showing possible variations in the total project cost, given the uncertainties that were included in the risk model. We can use this S-curve to determine

---

[*] (Hillson, 2014b)

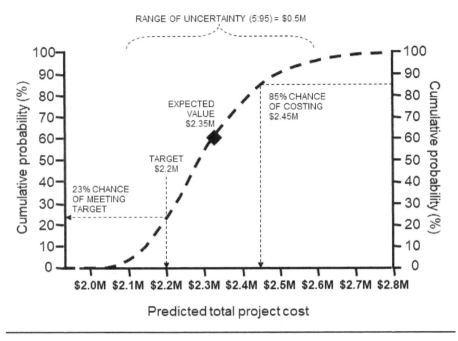

**Figure 7-9** Example S-Curve for Total Project Cost

the overall cost-risk exposure for this project by directly answering the two key quantitative questions.

- *How likely is this project to succeed (or fail)?* The S-curve in Figure 7-9 shows that the probability of meeting the project cost target of $2.2M is 23%, with a 77% chance of exceeding the budget. The analysis predicts an expected outcome of $2.35M, which is an overspend of $0.15M, or 7%. The project sponsor can determine values of total project cost that correspond to chosen confidence levels; for example, there would be an 85% chance of meeting a revised budget of $2.45M. This allows the project sponsor to make risk-informed decisions trading off increased cost (+ $0.25M) against increased probability of success (from 23% to 85%).
  In summary:
  ○ Probability of meeting $2.2M target = 23%
  ○ Expected value = $2.35M (+7%)

  This project has a 23% chance of succeeding in meeting its budget, and is likely to overspend by 7%, unless further action is taken to manage the current levels of uncertainty and risk.
- *What is the potential range of variation in outcome?* This example shows that the potential variation in total project cost is $0.5M against a target

budget of $2.2M (i.e., 22% of the expected project value), with a range of possible values from $2.1M (5th percentile) to $2.6M (95th percentile). This tells the project sponsor that in the best case they might expect to beat the budget by $0.1M (representing a 4% underspend), but at worst the project might exceed its budget by $0.4M (18% overrun).

Summarising these results:
- Total potential range = $0.5M (= 22% of project value)
- Realistic best case = $2.1M (–4%)
- Realistic worst case = $2.6M (+18%)

The project budget shows a potential variation of 22% (from –4% to +18%), unless cost risk is managed proactively during the remainder of the project.

The example in Figure 7-9 focuses only on the project budget, describing overall project risk solely in cost terms. It is of course possible to use QRA to evaluate overall project risk exposure against other project outcomes such as time, performance, return on investment (ROI), etc.

## MODIFYING THREAT TECHNIQUES TO MODEL OPPORTUNITIES

We've spent a lot of time in this chapter describing how QRA is usually done in projects in which the focus is only on threats. Other chapters have merely summarised the threat-based approach before turning to the opportunity side in more detail. There are two main reasons why this QRA chapter is different:

- Including opportunities in QRA models builds on the foundation of the threat-based approach. Unfortunately, QRA is usually not done well or properly, so it's important to explain how it should be done for threats before we can move on to opportunities.
- Once the proper way to model threats is understood, including opportunities is merely a simple extension of the basic techniques.

The main area in which threat-based QRA needs to be modified to include opportunities is in the definition of input data. Once positive uncertainty is included in the risk model as well as negative uncertainty, the standard QRA outputs can be interpreted in the same way.

In the earlier section, we described four ways in which threat-based input data should be generated in order to reflect all sources of uncertainty that might affect the project:

- *Variability.* This is modelled using different types of range estimates to represent potential variation in planned tasks or activities.
- *Ambiguity.* Here we reflect alternative futures that might occur by using stochastic branches, which can be either probabilistic or conditional.
- *Individual project risks.* Probabilistic branches can also be used to include the effect of a risk that might either occur or not.
- *Correlation.* The relationship between different types of uncertainty is modelled using dependency or correlation groups.

Each of these four approaches should also be used to include the effect of upside uncertainty in our risk model, so that it properly represents all sources of uncertainty. The use of ranges, branches, and correlation groups to model opportunities is explained in the following sections.

- *Variability.* This is modelled using *ranges of values* to reflect the fact that the actual outcomes for planned activities and tasks may be different from the plan. For example, a project activity may take more or less time to complete, and/or it may require more or fewer resources, and/or it may cost more or less. Figure 7-1 above illustrated several range types that can be used in a risk model to represent this variation. All of these ranges have a minimum value and a maximum value, which indicate the outside limits of expected variation, with the planned value somewhere between these two.

  Clearly, if the actual duration or cost or resource requirement were to turn out higher than what was planned, that would be undesirable, so the possibility of these higher values in the range actually happening is a threat. But conversely, the possibility that the actual outturn might be better than the original plan would represent an opportunity.

  Usually when QRA is used in an approach that only recognises downside risks, maximum values for ranges are related to such risks, but minimum values are more woolly and often the result of simple guesswork. In a risk approach that includes both opportunities and threats, it is natural to think about ways in which reality might turn out to be better than planned as well as worse, and the available range types in Figure 7-1 (and others) offer a natural way to model both upside and downside variation. When opportunities are explicitly included in the risk approach, developing minimum values for our ranges is no longer guesswork, but instead we have a rational basis for setting the lower limit.
- *Ambiguity.* Stochastic branches are used in risk models to represent the fact that there are several possible ways in which our project might develop in future, and we don't always know which one will happen. We can use

*probabilistic branches* to model a number of alternative paths for the project, each of which has an associated likelihood of occurring (see Figures 7-2 and 7-3 for examples). A *conditional branch* is used if the occurrence of a particular path depends on something else happening (see example in Figure 7-4).

Both of these types of stochastic branches are useful for modelling upside uncertainty, simply by including alternative paths for the project that are positive (to recognise that good things sometimes do happen on our projects!). For example, a cautious project manager might plan for the first acceptance trial to fail and a second to be needed, but in fact the repeat may be unnecessary. This is shown in Figure 7-10, which can be compared with Figure 7-2.

Similarly, the example branches shown in Figures 7-3 and 7-4 show alternative paths through the project that have different chances of occurring (Figure 7-3) or that only occur if certain conditions are met (Figure 7-4), and it is quite possible that one or more of these branches could be more favourable than the base plan.

- *Individual project risks (opportunities).* The earlier section suggested that the clearest way to model individual threats was to use a probabilistic branch, in which the probability associated with the branch is the chance that the threat will occur, and the content of the branch indicates the threat impact (see Figures 7-5 and 7-6). The same construct can be used for individual opportunities that have a probability of occurrence and a positive impact if they occur.

  The positive impact of an opportunity usually results in reduced time or cost or resources, and this can be simply modelled using a probabilistic branch as shown in Figures 7-11 and 7-12 (analogous to Figures 7-5 and 7-6). As for branches reflecting threats, the values of probability and impact for a probabilistic branch showing an opportunity are based on the information held in the risk register for that opportunity.

- *Correlation.* The use of correlation in risk models that include both upside and downside risks is exactly the same as its use in threat-only models. The only area that requires careful thought and attention is when we are correlating threats with opportunities, in which case we may need to use negative correlation coefficients: If a threat occurs, it might make an opportunity less likely or less beneficial or even impossible, and vice versa. Otherwise, the discussion in the earlier section on correlation is equally applicable here.

Having produced input data for our risk model that include both upside and downside uncertainties, we can then proceed to run a Monte Carlo simulation

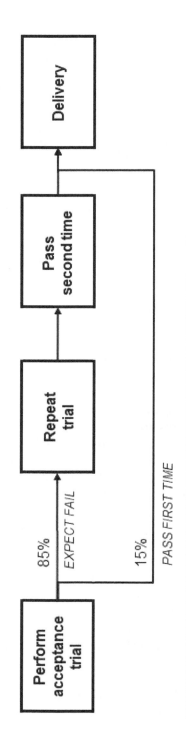

**Figure 7-10** Example Probabilistic Branch with Positive Alternative Path

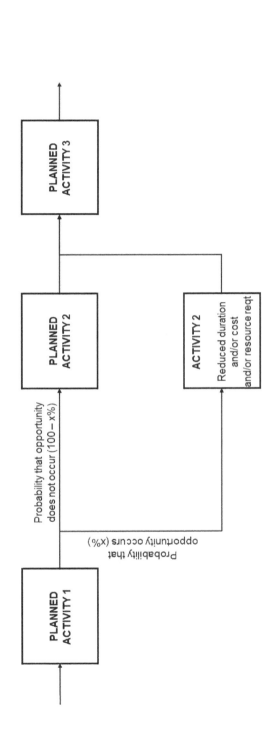

**Figure 7-11** Probabilistic Branch for Modelling an Opportunity into the Project Schedule

| COST ELEMENT 2.1 | ITEM COST |
|---|---|
| Cost item 2.1.1 | $ A |
| Cost item 2.1.2 | $ B |
| Cost item 2.1.3 | $ C |
| Optional opportunity cost item: | |
|     Probability that opportunity occurs (x%) | $ – D (saving) |
|     Probability that opportunity does not occur (100 – x%) | $ zero |
| TOTAL COST | |
|     With opportunity (x% chance of achieving this total) | $ A+B+C–D |
|     Without opportunity (100-x% chance of achieving this total) | $ A+B+C |

**Figure 7-12** Probabilistic Branch for Modelling an Opportunity into the Project Budget

in the normal way, producing the same set of outputs as usual, and making the same inferences about the range of possible project outcomes. The key difference when the dataset includes both threats and opportunities is that the analytical results will be more realistic, reflecting the full range of possible uncertainties that could affect the project, and this in turn will allow us to make better risk-informed decisions about our risk response strategies.

## WRITE IT DOWN

Unlike the preceding steps in the risk process, for which the resulting information was recorded in the project risk register, the results obtained from QRA cannot be included there. This is because entries in the risk register each contain information about one individual project risk, but QRA analyses the combined effect of all risks on overall project outcomes. Consequently, QRA results are for the whole project, and not per-risk.

It is, however, still important to record the findings of the QRA step, so that they can be used in subsequent phases of the risk process. Since the risk register is not suitable, we need to use a risk report to present QRA results. Risk report formats are discussed in more detail in Chapter 10, but the section covering QRA is likely to include some or all of the following elements:

- Purpose and scope of QRA on this project
- Modelling assumptions and limitations
- Sources of input data
- Model structure
- Key analytical outputs (S-curves, Tornado Charts, criticality analysis)
- Key analytical results

- Range of project outcomes, from best case to worst case, indicating the level of uncertainty
- Main drivers of uncertainty in project outcomes, including both individual project risks and planned activities/tasks
- Calculated expected values for project success measures, including project milestone dates, project duration and end dates, total project cost, cashflow, rate of return, profitability/margin, resource requirements, etc.
- Probability of achieving given project targets
- Confidence levels associated with particular outcome values
- Required levels of contingency
• Assessment of overall project risk exposure against project objectives

## SUMMARY AND REFLECTION QUESTIONS

Some people think that QRA is an optional extra, and most projects choose to opt out. For small or simple projects, this may be appropriate, if sufficient information is gathered during the qualitative risk assessment step to support effective risk response planning. For larger or more complex projects, a deeper analysis of the effects of uncertainty on the project will usually be needed.

However, whatever the project size, there's only one way to answer the vital question, *"How risky is this project?"* The answer has to take account of all forms of uncertainty, including individual threats and opportunities of course, but also sources of variability and ambiguity in the project. Combining all these disparate types of uncertainty requires use of computer models and simulation techniques, typically involving Monte Carlo simulation. It's not possible to answer the question without **using numbers** to represent uncertainty.

In order to be realistic, it's vital that QRA simulation is not limited to negative or downside risk. Risk models that only include threats can only be pessimistic, because they have no way of reflecting upside impacts. A realistic QRA has to take account of individual opportunities, as well as upside types of variability and ambiguity. Only then can we be sure that we have a proper view of the overall risk exposure of our project, allowing us to make sound, risk-informed decisions about the future.

### Reflection questions

- What is "overall project risk", and how can QRA be used to evaluate it?
- What are the most common shortcomings when projects use QRA?

- Why is a risk model that excludes upside uncertainty both useless and misleading?!
- Explain the difference between variability, ambiguity, and individual project risks, and give an example of each. How can these be represented in a QRA risk model?
- When would you use each of the range types in Figure 7-1?
- Why is it essential to include correlation (or dependency) in a risk model? What happens to the results if you don't?
- Which is worse: using a flawed risk model for QRA, or not using QRA at all?

## THE NEXT STEP ("NOW WE'VE QUANTIFIED OVERALL RISK EXPOSURE, WHAT CAN WE DO ABOUT IT?")

So far in the risk process, we've clarified our objectives and set risk thresholds, identified and recorded individual threats and opportunities, then prioritised and grouped these risks to find the worst threats and the best opportunities. We may also have used quantitative risk analysis to assess the combined effect of all our threats and opportunities, together with other sources of uncertainty, on project outcomes, allowing us to evaluate overall project risk exposure.

After all this analysis, surely it must be time now to actually *do* something?! When do we stop analysing risks and start acting to manage them?

Well, almost . . . The temptation is to leap into action, doing whatever we can to reduce our overall risk exposure. But there's one vital step before we can act. We need to decide what to do—and that's the topic of our next chapter.

# Chapter 8

# Deciding What to Do

Previous steps in the risk management process have concentrated on identifying, understanding, and analysing the threats and opportunities in our project. These steps are clearly vital, because we can't manage risks which are not identified, or which are poorly understood or even misunderstood. But we mustn't stop there. If we don't do anything with the risk information we've gathered, we gain no benefit. Diagnosis is the not the same as cure, and *analysis* must lead to *action*. In most cases, simply identifying and understanding a risk does not make it go away. Unless the risk process results in action, risks remain unmanaged, and the process is largely a waste of time. We need to **decide what to do**. This chapter describes our options, and explains how to choose the right course of action for each individual risk, and for the level of overall project risk exposure.

## PURPOSE AND PRINCIPLES OF RISK RESPONSE PLANNING

It's easy to see why people think that this phase of the risk process is the most important (although, as we'll see in Chapter 9, that's probably not correct!). The effectiveness of our risk responses will directly determine whether the risk exposure of the project gets better or worse. If we get this part right, then threats will be reduced in probability and/or impact, and some will be avoided altogether, meaning that our project will suffer from fewer problems and issues, and the ones that do occur will be less severe. In addition, with effective risk responses we'll be able to capture more of the available opportunities, turning them into savings for the project, and we might also improve other opportunities so that they offer us even more benefit, helping us to work faster/smarter/cheaper and giving us the best chance to deliver a successful project. We can also manage overall project risk exposure to keep it within the risk thresholds set for our project.

On the other hand, if our risk responses are ineffective, threats and opportunities will go unmanaged, leading to avoidable problems and missed benefits. And poor risk response planning can make things even worse than we started with, introducing new threats or closing off possible opportunities, and allowing overall project risk exposure to go beyond the acceptable risk thresholds.

And if we have no way of knowing whether our planned risk responses are good or not, then we might proceed with our project falsely believing that all our risks are well managed and we'll be OK, when the opposite is true, leaving us unprepared for what might happen in the future.

This is why it's essential that we pay proper attention to risk response planning. This is where we can really make a difference to our project risk exposure—or not. So ***the purpose of risk response planning is to identify appropriate ways to address individual threats and opportunities, as well as ways to manage overall project risk.***

The following ***principles*** are important to remember if we want to plan effective risk responses.

## Strategy before tactics

When project teams start to think about risk responses, it's common to leap straight into detailed action planning without knowing what we're trying to achieve. This can result in a lot of activity but a lack of effectiveness, especially if actions are not well coordinated. To overcome this, a two-stage approach should be followed: First define the appropriate strategy for each risk, or to address the level of overall project risk; then design tactics to implement the chosen strategy.

It is important to determine the appropriate strategy first, then to develop actions to put it into practice. There is no single "best" risk response strategy, but the number of available strategy options is limited, making it easier to decide among them.

Having selected the appropriate strategy, attention can then be given to development of tactical actions which aim to realise the strategy. This creates focus in risk response planning, and avoids the "scatter-gun" approach, in which a number of alternative actions may be considered, some of which may negate the effect of others. Determining strategy first will ensure that actions are aiming for the same goal, and should avoid nugatory effort.

## Deal equally with both threats and opportunities

It may seem natural to focus first on addressing threats that could adversely affect our ability to achieve project objectives, and leave opportunities until later. There are several problems with this approach, not least of which is that we're quite likely to spend so much time and effort on threats that we don't get around to doing anything about opportunities. In addition, planning responses first for threats before we turn to opportunities means that we'll be considering lower-priority threats ahead of high-priority opportunities. In other words, we're ignoring the results of the qualitative risk assessment phase, when we prioritised our risks to find the worst threats and the best opportunities.

Although there may be strong psychological drivers that influence us to deal first with threats (remember Maslow?[*]), a key principle of this step of the risk process is to deal equally with both threats and opportunities. This means treating high-priority opportunities with as much attention as high-priority threats.

## Respond to overall risk, not just individual project risks

One important output from the quantitative risk analysis phase is an evaluation of overall project risk exposure. We should respond to this information appropriately, planning risk responses that will keep overall risk within acceptable thresholds. Too often, risk response planning ignores overall project risk and only considers individual threats and opportunities.

## The first idea is not always best

When we've got a long list of risks in the risk register, each of which needs an appropriate response, there's a tendency to go with the first thing we think of, and move swiftly on to the next risk. But there's no guarantee that our first idea will be the best. In principle, we should always consider alternatives and options, and then choose the one that gives us the required outcome.

In practice, the way to address some risks is immediately clear, and we don't need to spend a lot of time on them. This might be particularly true of low-priority risks, which have a small chance of occurring or whose impact would be insignificant. But high-priority risks should be given more careful attention.

Each alternative risk response will have an associated cost as well as a potential effect on the risk. In order to choose the "best" response, we need to ensure cost-effectiveness as well as appropriateness. Do we really need an aggressive

---

[*] (Maslow, 1943, 1987)

response if it would be very expensive? Might it be better to take a less expensive approach that gives an acceptable result, rather than going all out?

## Responses must match the level of risk

We need to be sure that our planned responses are *appropriate*. Not all risks demand the same intensity of response. For some risks, it's entirely appropriate to panic, if they require an urgent or high-level response. This might include an emergent threat that demands a crisis response or the project cannot proceed, or an unmissable opportunity that needs immediate action or it will pass. Alternatively, for some risks it might be appropriate for us to do nothing, except maybe monitor the risk in case it changes. In some cases, it could be appropriate to stop the project until a particular risk has been dealt with, whereas other risks can be completely ignored.

Although the appropriate response for most risks will lie somewhere between panic and passivity, it's vital to get the level of response right, so that major threats or opportunities are not ignored while the project wastes valuable time and resources on tackling minor ones.

## Be creative but realistic

There are two parts of the risk management process in which we need to be creative. The first is when we identify risks, and the second is when we develop appropriate responses to them. The reason we need creative thinking is that we're trying to imagine uncertain things that may not have happened before, and we're aiming to develop actions that we hadn't previously realised would be necessary in our project. Both of these challenges require innovative, out-of-the-box thinking, and we should use techniques that stimulate creativity in both risk identification and risk response planning.

However, creativity needs boundaries. If we allow unfettered imaginations to run riot, we can waste a lot of time considering "risks" that are so unlikely or insignificant as to be irrelevant. We can also be distracted by crazy risk response proposals that could never be implemented.

All proposed risk responses must be *achievable*. There is no point in describing risk responses which are not realistic or feasible, either technically or within the scope of our capability and responsibility. If your planned response is "Hope for a miracle" or "Invent a radical new solution", you may be disappointed! You might decide that we should double the size of our testing team, but if the organisation has imposed a recruitment freeze then you can't do it. Yes, it might be helpful to employ the world's leading technical expert on our team, but it's not going to happen.

Responses must also be *affordable*. The amount of time, effort, and money spent on addressing the risk can't exceed the available budget or the degree of risk exposure. Each risk response should have an agreed budget, added to the approved project cost plan.

## Ensure clear ownership

Many times, a project team will develop good risk responses in a workshop, but fail to gain agreement from the people who are expected to implement the agreed responses. Each risk response strategy should be owned by a single person (and accepted by them) to ensure a single point of responsibility and accountability for implementing the response. This requires careful delegation, including provision of the necessary resources and support to allow effective action to be taken.

The person who agrees to take responsibility for managing a particular risk is often called the *risk owner*. However, they may not be best placed to actually take the necessary actions to implement the agreed risk response strategy. In this case, the risk owner may involve one or more *action owners* who will perform the required tasks.

## Summary

The purpose and principles of risk response planning are summarised in Table 8-1.

Table 8-1 Purpose and Principles of Risk Response Planning

| | |
|---|---|
| Purpose | To identify appropriate ways to address individual threats and opportunities, as well as ways to manage overall project risk |
| Principles | • Strategy before tactics<br>• Deal equally with both threats and opportunities<br>• Respond to overall risk, not just individual project risks<br>• The first idea is not always best<br>• Responses must match the level of risk<br>• Be creative but realistic<br>• Ensure clear ownership |

## TYPICAL TECHNIQUES FOR RESPONDING TO THREATS

Risk response strategies for individual threats are well established,* and usually fall under the following five headings:

- **Escalate.** Threats for which the impact is outside the scope of our project should be passed to the person or party whose objectives would be affected if the risk occurred. Because the risk would not affect our project, once it has been escalated we do not need to record it in our project risk register.
- **Avoid.** The aim of avoidance is to eliminate the risk, either by making the threat impossible or by protecting the project against its potential impact.
- **Transfer.** This requires involving another person or party in managing the risk, giving responsibility to someone who is better able to manage the risk than the project team. The risk remains on the project risk register because it has not gone away, but responsibility for its management has been delegated to a third party. This may involve payment of a fee, or placement of a contract.
- **Reduce.** Reduction of a threat aims to reduce its probability and/or impact, ideally bringing it below a risk threshold at which it becomes acceptable to the project.
- **Accept.** For residual threats for which proactive action is either not possible or not cost-effective, acceptance is the last resort, taking the risk without special action. A contingency plan is often developed, to be implemented if the threat actually occurs.

These strategy types are usually only considered as relating to individual threats within a project, but they are also applicable to address the level of overall project risk (with the exception of Escalate). For example:

- The **Avoid** strategy might lead to project cancellation if the overall level of risk remains unacceptable.
- **Transfer** at overall project level could result in setting up a collaborative business structure in which the customer and the supplier share the risk.
- **Reduction** of overall project risk exposure can be achieved by re-planning the project or changing its scope and boundaries to remove risky elements.
- **Accepting** an overall level of negative project risk means that the project will be continued without significant change, although the organisation may make contingency plans and monitor exposure against predefined trigger conditions.

---

* (Hillson, 1999; Project Management Institute, 2009, 2017)

# MODIFYING THREAT TECHNIQUES TO RESPOND TO OPPORTUNITIES

Because many project teams are used to the five common risk response strategies for threats of *escalate, avoid, transfer, reduce,* and *accept*, it seems sensible to build on these as a foundation for developing strategies appropriate for responding to identified opportunities.[*] We can do this by generalising the underlying principle behind each threat strategy, then finding the positive equivalent approach for dealing with opportunities. The principle is illustrated in Table 8-2, and detailed in the paragraphs below.

Table 8-2 Generalising Threat Response Strategies to Deal with Opportunities

| Threat Response | Generic Strategy | Opportunity Response |
|---|---|---|
| Escalate | Pass risk to relevant owner | Escalate |
| Avoid | Eliminate uncertainty | Exploit |
| Transfer | Allocate responsibility | Share |
| Reduce | Modify exposure | Enhance |
| Accept | Include in baseline | Accept |

Generalising and extending the five common threat strategies allows us to produce opportunity-focused equivalents, as follows:

- **Escalate** is used for a threat that lies outside the project scope, and involves *passing it to the relevant owner* of the objective that would be affected. The same approach can be adopted to **escalate** opportunities whose effect is outside the project.
- **Avoidance** strategies which seek to remove threats are actually aiming to *eliminate uncertainty*. The upside equivalent is to **exploit** identified opportunities—removing the uncertainty by seeking to make the opportunity definitely happen.
- **Risk transfer** is about *allocating responsibility* to enable effective management of a threat. This can be mirrored by **sharing** opportunities—passing ownership to a third party best able to manage the opportunity and maximise the chance of it happening.
- **Reduction** seeks to *modify the degree of risk exposure*, and for threats this involves making the probability and/or impact smaller. The opportunity

---

[*] (Hillson 2001, 2002g)

equivalent is to **enhance** the opportunity, increasing its probability and/or impact to maximise the benefit to the project.
- The **accept** response to threats *includes the residual risk in the baseline* with no special measures apart from development of a contingency plan. Opportunities included in the baseline can similarly be **accepted**—adopting a reactive approach without taking explicit actions, apart from developing a suitable contingency plan to be implemented if the opportunity occurs.

Let's look in more detail at each of these opportunity strategies.

## Escalate

The **escalate** strategy for opportunities is essentially identical to the threat version. When we find an opportunity that would not affect our project objectives if it happened, but that could affect another project or another part of the organisation, we might choose to ignore it and hope that the relevant people also discover it in time. Instead of this somewhat irresponsible and blinkered approach, we should pass details of the opportunity to the person or party whose objectives would be affected if the opportunity occurred. Because it cannot affect our project, an escalated opportunity does not need to be recorded in our project risk register.

## Exploit

The aim of this risk response strategy is to eliminate the uncertainty associated with a particular opportunity. **Exploit** seeks to do this by making the opportunity definitely happen. Whereas the threat equivalent strategy of **avoid** aims to reduce probability of occurrence to zero, the goal of the **exploit** strategy for opportunities is to raise the probability to 100%; in both cases the uncertainty is removed. This is the most aggressive of the response strategies, and is usually applied to "golden opportunities" with high probability and potentially high positive impact, which the project or organisation cannot afford to miss.

In the same way that **avoidance** for threats can be achieved either directly or indirectly, there are also direct and indirect approaches for **exploiting** opportunities. Direct responses include making positive decisions to include an opportunity in the project scope or baseline, removing the uncertainty over whether or not it might be achieved by ensuring that the potential opportunity is definitely locked into the project, rather than leaving it to chance. Indirect exploitation responses involve doing the project in a different way in order to allow the opportunity to be achieved while still meeting the project objectives—for example, by changing the selected methodology or technology. In the same way as **avoidance** goes around a threat so that it cannot affect the project, **exploitation** stands in the way of the opportunity to make sure that it is not missed, in effect making it unavoidable.

## Share

One objective of risk response planning is to ensure that ownership of the risk response is allocated to the person or party best able to manage the risk effectively. For a threat, **transferring** it passes responsibility for its management to a third party. Similarly, **sharing** an opportunity involves allocating ownership to a third party who is best able to handle it, in terms of maximising the probability of occurrence, and/or increasing potential benefits should the opportunity occur.

Clearly it is sensible to consider project stakeholders as potential owners of this type of response, because they already have a declared vested interest in the project, and are therefore likely to be prepared to take responsibility for managing identified opportunities proactively.

A number of contractual mechanisms can be used to **transfer** threats between different parties, and similar approaches can be used for **sharing** opportunities. Risk-sharing partnerships, special-purpose companies, or joint ventures can be established with the express purpose of managing opportunities. The risk-reward arrangements in such situations must ensure equitable division of the benefits arising from any opportunities that may be realised. The target-cost-incentivisation type of contract is also suitable for both threats and opportunities, because it provides a mechanism for distributing either profit or loss.

It is important that **sharing** of a project opportunity does not become mere abdication of responsibility on the part of the project manager, who should retain an active involvement in the management of all risks which could affect project objectives.

## Enhance

For risks which cannot be escalated, avoided/exploited, or transferred/shared, the fourth type of response strategy aims to modify the "size" of the risk to make it more acceptable. In the case of threats, the aim is to **reduce** the probability of occurrence and/or severity of impact on project objectives. In the same way, opportunities can be **enhanced** by increasing probability and/or impact.

The probability of an opportunity occurring might be increased by seeking to facilitate or strengthen the cause of the risk, proactively targeting and reinforcing any trigger conditions that may have been identified. Impact drivers which influence the extent of the positive effect can also be targeted, seeking to increase the project's susceptibility to the opportunity, and hence maximise the benefits should it occur.

Where several opportunities have been identified as arising from a common cause, it may be particularly cost-effective to look for generic **enhancement** actions which target the common cause. If these actions are successful they will influence more than one opportunity, and could result in a significant increase in benefits to the project.

## Accept

Residual risks are those which remain after other response options have been exhausted. They also include those minor risks for which any response is not likely to be cost-effective, as well as uncontrollable risks for which positive action is not possible. The common terminology adopted for threats in these categories is to **accept** the risk, with application of contingency where appropriate, and ongoing reviews to monitor and control risk exposure.

Opportunities which cannot be actively addressed through escalating, exploiting, sharing, or enhancing can also be **accepted**, with no special measures being taken to address them. In the same way as accepting threats, we can accept opportunities by doing nothing and hoping to "get lucky". For a threat this would mean hoping that it will not occur, but for an opportunity we hope that it will.

One way in which opportunities can be included in the project baseline without taking special action to address them is by appropriate contingency planning. As for threats, this involves determining what actions will be taken should the opportunity occur, preparing plans to be implemented in that eventuality. Funds could be set aside to be spent on emerging opportunities, or resources and facilities nominated to be used if necessary.

It is also important for the project team to remain risk-aware, monitoring the status of identified opportunities alongside threats to ensure that no unexpected changes arise, and the use of an integrated risk process to manage both threats and opportunities together will assist in achieving this goal.

Threat-focused risk response strategies can be applied to tackle overall project risk in which the level of risk exposure is unacceptably negative (apart from Escalate, which is not relevant at overall project level). In the same way, we can use our opportunity response strategies at overall project level if this is particularly positive. Examples include the following:

- **Exploiting** positive overall project risk could result in an agreed expansion of the project scope, or the launch of a new project.
- **Share** might be implemented through joint ventures or special-purpose vehicles in which parties divide any benefits arising from upside risk exposure, often using an agreed gain-share formula.
- The **Enhance** strategy can be achieved at the overall project level by re-planning the project or changing its scope and boundaries to take advantage of the upside exposure and incorporate more risky elements.
- **Accepting** an overall level of positive project risk means that the project will be continued without significant change, although the organisation may make contingency plans and monitor exposure against predefined trigger conditions.

## SELECTING PREFERRED STRATEGY

One of the principles of risk response planning is "strategy before tactics". Another principle is that the first idea may not always be the best, so we should always consider alternatives and options. This means we need a framework for comparing risk strategies and deciding which one to follow, before we move to detailed action planning.

The only exception to the need to consider several options is for risks that are appropriate to be escalated. These are risks whose effect is outside the scope of the project, and that would not impact a project objective if they occurred. However, if these risks would affect someone else's objective, then we clearly need to pass the risk on to that person so that they can manage it. In these cases, escalate is the only option, and no other strategy should be considered, since the risk is not ours to manage.

For other risks that are not escalated, we need to select a preferred risk response strategy, one that will shape our action planning in order to actually manage the risk. Of course, the strategy that we select may turn out not to be as effective as we had hoped, so we must always be prepared to change direction later in the project if necessary. But the point of choosing a single risk response strategy for each risk is to allow us to focus our risk action planning, and this can only work if we have a single strategy at any one time. This must be *the best strategy for now*, recognising that another risk strategy may be better later.

One possibility is to use the position of the risk in the Probability-Impact Matrix (P-I Matrix) as an indicator of the preferred response strategy, as shown in Figure 8-1. This suggests that risks with a high probability of occurrence and high impact require aggressive treatment, and should be *avoided* if possible if they pose a significant threat to the project, or *exploited* if they represent a major potential opportunity. Those minor risks with low probability and low

170  Capturing Upside Risk: Finding and Managing Opportunities in Projects

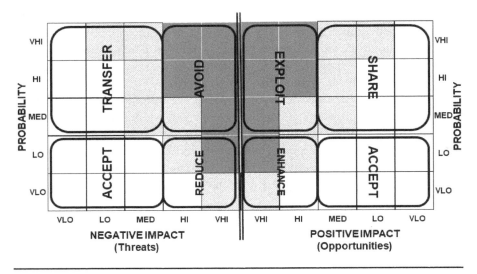

**Figure 8-1**  Selecting Risk Response Strategy by Position on P-I Matrix

impact can be *accepted*, whether they are threats or opportunities, since they are unlikely to occur, and would not affect the project significantly if they did happen. High probability/low impact risks might be considered as safe candidates for passing to another stakeholder via *transfer* or *share* responses, because the penalty to the project would be small if the recipient failed to manage the risk effectively, either in realised negative impact or in missed benefits. Finally, threats with low probability but potentially high adverse impacts should be *reduced* if possible, and we should look for ways to *enhance* opportunities with low probability but potentially high beneficial impacts.

This might appear overly simplistic, because cases could be made for considering different strategies for risks in each quadrant. An alternative approach might be to consider risk response strategies in order of their potential effect on the risk. Under this approach, *avoid* would be seen as the first-choice strategy for a threat, and *exploit* for an opportunity, because these are the most aggressive strategies and would result in elimination of the uncertainty—avoiding a threat means it can't happen, and exploiting an opportunity means it will happen. The second priority strategy should be to consider whether some other party might be better able to manage the risk than the project or organisation itself, seeking another stakeholder who would be prepared to take the risk on, by either *transferring* a threat or *sharing* an opportunity. Where it is not possible or cost-effective to eliminate the risk or to pass ownership to another party, we should try to develop responses which *reduce* threats and *enhance* opportunities. The last-resort option is to *accept* residual risks which remain after other

**Figure 8-2**  Prioritising Risk Response Strategies by Intensity

strategies have been considered, or where proactive strategies are not practicable or cost-effective. This gives us a priority order for considering risk response strategies, based on the intensity of their possible effect on each risk, as shown in Figure 8-2.

Both of these frameworks for selecting a preferred risk response strategy are reasonable starting places, but they are rather mechanistic. A particular risk response strategy may not be appropriate for all risks that are mapped into a given quadrant of the P-I Matrix. Just because we *can* adopt an aggressive risk response strategy (such as avoid or exploit) doesn't mean we *should*. We need to consider four other factors before choosing a preferred risk response strategy:

- Cost-effectiveness
- Risk-effectiveness
- Timeliness
- Secondary risks

## Cost-effectiveness

Implementing risk responses is usually not free. Each response is likely to involve expenditure of additional time, cost, or resource. Clearly it is important that the organisation should be prepared to spend the required time, money, or effort in responding to identified risks, otherwise the process will be ineffective. However, we need assurance that spend now is justified in order to remove exposure later. We also need to be sure that the amount of expenditure is appropriate to the size of risk faced. For example, we don't want to spend $100,000 on a response to a risk whose maximum impact might be $10,000 (unless there are other impacts such as company reputation, safety or environmental implications, or "time is of the essence" considerations).

In some cases, we can perform a cost-benefit analysis for proposed responses to determine if they're cost-effective. This only works for risks (threats or opportunities) for which the impact is financial, or if we can convert other types of impact into monetary terms (including, for example, the cost of delay, or the cost of rectifying performance shortfalls). We must also be able to assess the cost impact of the risk before we implement the response, as well as the cost impact that we would expect if the risk response is successful. For threats, we want the post-response cost impact to be less than the current cost impact by as much as possible. For opportunities, the opposite is true: We want to increase the potential positive savings available if the opportunity occurs, and bigger savings are better.

Then we can calculate a cost-benefit ratio (CBR) for risk response strategies that target either threats or opportunities, as shown in Figure 8-3.

$$CBR_{threat} = \frac{[Cost\ impact]_{before\ response} - [Cost\ impact]_{after\ response}}{[Cost\ of\ response]}$$

$$CBR_{opportunity} = \frac{[Cost\ impact]_{after\ response} - [Cost\ impact]_{before\ response}}{[Cost\ of\ response]}$$

**Figure 8-3** Cost-Benefit Ratio Calculations for Threats and Opportunities

This gives the ratio of the improvement in cost-risk exposure to the cost of obtaining that improvement. The larger the CBR, the more cost-effective the response. Values of CBR less than 1 cost more now than they might save later. As a guideline, effective responses should have values of CBR above 20. CBR can be used to compare alternative proposed responses, allowing the most cost-effective risk response strategy to be selected.

## Risk-effectiveness

Calculation of CBR is only possible if all impacts of a risk can be converted into money, and if the "before and after" cost impacts can be estimated accurately. For risks where this is not possible, we can still perform a type of cost-benefit analysis, but the benefit is expressed as a change in risk score.

The use of a risk score for a risk based on estimates of its probability and impact was discussed in Chapter 6, and an example risk scoring scheme is presented in Figure 6-3. This involves calculation of a risk score for each risk regardless of its impact type, allowing us to prioritise risks on a common basis of comparison. These risk scores can also be used to perform a cost-benefit analysis for risk response strategies, by estimating changes in risk score that would be expected if we implemented a particular risk response, and comparing that with the cost of the response. In the same way that we can determine the cost-effectiveness of risk responses by calculating CBR for risks with financial impacts, we can estimate the risk-effectiveness of proposed responses by calculating a generic risk-benefit ratio (RBR). For threats, we want to see the biggest possible reduction in risk score for the minimum cost, and for opportunities we're seeking to increase risk score.

RBR is calculated as shown in Figure 8-4.

$$RBR_{threat} = \frac{[\text{Risk score}]_{\text{before response}} - [\text{Risk score}]_{\text{after response}}}{[\text{Cost of response}]}$$

$$RBR_{opportunity} = \frac{[\text{Risk score}]_{\text{after response}} - [\text{Risk score}]_{\text{before response}}}{[\text{Cost of response}]}$$

**Figure 8-4** Risk-Benefit Ratio Calculations for Threats and Opportunities

As for CBR, bigger RBR values are better, because they represent more cost-effective reduction of threats or enhancement of opportunities. We can therefore use RBR to compare alternative risk response strategies, allowing us to select the one that gives us the biggest bang-per-buck.

## Timeliness

An additional factor to be considered when selecting the preferred risk response strategy is *timing*. In Chapter 6, as part of the qualitative risk assessment process, we discussed the importance of timing factors when prioritising risks. In particular, we introduced proximity and urgency:

- *Proximity* describes how soon this risk is expected to occur, if it does occur, and this is known as the "impact window". If a risk might happen soon (high proximity), then we need to prioritise it for action, if possible.

- *Urgency* describes how soon responses must be implemented in order to address this risk effectively, called the "action window". If we need to act quickly (high urgency), then this is a high-priority risk.

A risk with high proximity and high urgency should clearly have the highest priority when we are choosing a preferred risk response strategy—it is due to happen soon, and we have to act quickly if we want to manage it. Here we need to understand the speed with which we are able to implement each risk response strategy, and compare this with the impact window and the action window for the risk (see Figure 6-7). We may have identified a possible response that would address the risk effectively, but if we can't implement it in time then it is of no use and we must reject it and look for another response option.

## Secondary risks

We want our chosen risk response strategy to change the risk profile of the project in a positive way, but we can't assume that it will work as planned. The law of unintended consequences applies to risk responses in the same way as elsewhere. Sometimes implementation of a response may introduce more risk into the project than it removes.

Risks which arise as a direct result of implementing a response are called *secondary risks*. These should be identified and assessed in the same way as any other risks, and we need to consider the introduction of secondary risks as part of our evaluation of possible risk response strategies.

In a risk approach that only considers threats, it's easy to imagine a response to one threat introducing another. For example, if our project faces a threat of inability to meet the client's requirement as a result of lack of technical expertise, and we decide to subcontract an element of our solution, we introduce a new threat of non-performance by our chosen subcontractor. That secondary threat only exists because we implemented the subcontracting response to our original threat.

However, when our risk process includes both threats and opportunities, the range of potential secondary threats becomes wider, with four alternative situations to consider:

- The worst case is one in which a response to a threat introduces another secondary threat, making the situation worse than before.
- The comparable situation with opportunities occurs when a planned response to deal with one opportunity may have a double benefit and introduce a secondary additional opportunity.

- It is also possible that a response designed to address a threat not only achieves its aim, but also introduces an additional upside—i.e., the secondary risk is an opportunity. In this case there might be two beneficial results, with the original threat being reduced or eliminated and a new opportunity being created.
- Similarly, when a response is planned to tackle an opportunity, it may result in a secondary threat. Here the potential benefit of dealing with the opportunity might be discounted by the introduction of a new threat, and the balance must be considered carefully.

Despite the potential complications, secondary risks can be significant in our selection of the preferred risk response strategy, and their possible existence cannot be ignored.

## ENSURING OWNERSHIP

After we've selected a preferred risk response strategy for each individual risk and for the overall level of project risk, we need to identify a person who will take responsibility for ensuring that the strategy is implemented. This person is known as the *risk owner*.

For overall project risk exposure, the risk owner is obviously the project manager. He or she is responsible for ensuring that overall project risk remains within the set thresholds, and must take the necessary action, involving the project team, the project sponsor, and other stakeholders as needed.

But it is often less clear who the risk owner for individual threats and opportunities should be.

Fortunately, our initial definition of risk as "uncertainty that matters" can help us here. Every risk is defined in terms of the objectives that would be affected if the risk occurred. We can determine who will be the right risk owner for each risk by considering which objective, or subobjective, is impacted by the risk. The risk owner is the person who is responsible for delivering the affected objective/subobjective. If risk is "uncertainty that matters", find out who it matters to, and they are likely to be the right risk owner.

Although each risk will have a single risk owner, it is unlikely that this person will be able to manage the risk on their own. They will often need the assistance of others with specific skills, knowledge, or expertise. The risk owner should be responsible for turning the selected risk response strategy into specific actions, which will achieve the strategy if they are completed successfully. Where necessary, the risk owner should allocate actions to named *action owners*, who are accountable to the risk owner for ensuring that their action is completed within the specified parameters.

## TURNING STRATEGY INTO ACTIONS

Strategy changes nothing—it merely provides a statement of intent. The only value of strategy is in providing a focus for development of tactical actions that will implement the chosen strategy effectively, if the actions are themselves performed well. As a result, the risk response planning step doesn't end when we've selected a preferred risk response strategy for each individual threat and opportunity, as well as for the current level of overall project risk exposure. We need to follow through and turn strategy into actions.

Each individual risk is allocated to a single risk owner, who takes responsibility for ensuring that the selected risk response strategy is followed. This requires detailed action planning: developing a series of specific actions which, when implemented, will result in the strategy being achieved. Each action needs to be properly specified in sufficient detail to ensure that the person tasked with performing the action will know exactly what they have to do. This means that every action should have the following:

- Clear description of the task and deliverables
- Specified completion criteria
- Agreed duration, cost, and resources for the action
- Nominated action owner

Ideally, risk actions will be added into the project plan and treated like any other project task. Sometimes this is not possible—for example, if the project plan is a configurable controlled item—in which case risk actions should be documented in work allocation documents or task forms.

## WRITE IT DOWN

Developing risk responses generates additional data for each individual risk, which need to be recorded in the project risk register. The data fields to be completed in the risk register following risk response planning are shown in Table 8-3, for a minimalist, typical, or detailed risk register format.

Following the risk response planning step, we also now know what we need to do to manage the level of overall project risk, and we should write this down too. But this information doesn't belong in the project risk register, because it is at a higher level than individual risks. Chapter 7 explains the use of quantitative risk analysis (QRA) to evaluate overall project risk, and the results are recorded in a risk report. This report is also the natural place for us to record the strategy and actions we are planning to take in order to respond to overall project risk exposure.

Table 8-3  Risk Data to Record after Risk Response Planning

| Data Field | Risk Register Level | | |
| --- | --- | --- | --- |
| | Minimum | Typical | Detailed |
| **Risk Response Data** | | | |
| Risk owner | X | X | X |
| Preferred risk strategy | X | X | X |
| Summary of risk actions:<br>• Action owner<br>• Task description<br>• Action status | X | X | |
| Detailed risk actions including for each action:<br>• Action owner<br>• Task description<br>• Completion criteria/deliverables<br>• Action window<br>• Duration/budget/resource requirement<br>• Action status | | | X |
| Predicted probability/frequency of occurrence (post-response)<br>• Qualitative (for example, High, Medium, Low)<br>• Quantitative (for example, % range, frequency) | X | X | X<br>X |
| Predicted impact on each project objective (post-response)<br>• Qualitative (for example, High, Medium, Low)<br>• Quantitative (for example, three-point-estimates of days, dollars, technical performance measures) | X | X | X<br>X |
| Predicted overall risk ranking<br>• Red/Yellow/Green (or similar)<br>• Risk Score (calculated from probability and impact) | X | X<br>X | X<br>X |
| Secondary risks (if any) | | | X |

## SUMMARY AND REFLECTION QUESTIONS

Having identified and prioritised individual risks and evaluated the level of overall project risk exposure, it's vital that we then decide what to do. We know which are the worst threats and the best opportunities, and we need to develop effective actions to deal appropriately with them. By extending the familiar threat risk response strategies, we have comparable strategies to address opportunities, giving us a comprehensive set of options with which to tackle the risks facing the project. We can select a preferred strategy for each individual risk, in order to focus our attention and action, and then turn that strategy into specific actions to ensure that threats are avoided or minimised and opportunities are exploited or maximised. A single risk owner will be responsible for management of each risk, with the help of action owners where necessary. We've also chosen a risk response strategy to address the level of overall project risk exposure, and the project manager will take the lead on making that happen, with assistance from the project team, project sponsor, and other stakeholders as required.

**Reflection questions**

- List the five possible risk response strategies for threats and the matching five strategies for opportunities. Give an example of each from your current project.
- How can you choose the best risk response strategy for each risk?
- Why is it important to select a preferred risk response strategy before planning detailed actions? What's wrong with just deciding what to do?
- Many projects only develop risk responses for threats, and don't consider opportunities. Why is this a bad idea?

## THE NEXT STEP ("NOW WE'VE PLANNED RESPONSES TO OUR OPPORTUNITIES, LET'S MAKE SURE WE DO THEM.")

Everything is now in place for us to actually manage risk! We know what we're going to do and who's going to do it. But we still haven't done anything. If we don't turn analysis into action, all our previous effort will have been wasted. It's time to turn to the most important part of the risk management process, and act.

# Chapter 9
# Taking Action

Everything in the risk process up to this point has been data-gathering and analysis, and we've not actually done anything yet to manage the risks. We've identified the risks, prioritised them, perhaps modelled their combined effect on the overall project, and planned some responses. But if we don't do anything, nothing changes.

We've reached the part of the risk management process at which we get to actually manage threats and opportunities, and change the level of risk exposure on our project. If we don't do this, then everything else will be a waste of time, just working out what the risk challenge is, and considering what we could do, but not doing it. It's time to **take action**, as described in this chapter.

This chapter is much shorter than those describing other steps in the risk process, because the actions to be taken are largely common sense—we just have to do what we said we'd do. The reader may therefore be tempted to skip this chapter, thinking that its brevity indicates that it is less important. This mirrors an equal error made by many when they are trying to manage risk on their projects. The need to implement agreed risk responses is so obvious that it doesn't need any attention. Surely it will just happen, won't it? We don't need a separate step in the risk process, do we? Experience tells a different story.

## PURPOSE AND PRINCIPLES OF RISK RESPONSE IMPLEMENTATION

People used to think that everything up to this stage in the risk process is specifically about risk, but at this point, once we've decided what to do, the focus of action should pass out of the risk process and into routine project management, where risk actions will be implemented. In practice, people often don't implement agreed risk responses because they're seen as extra activities that people don't have time for, and that aren't as important as "real project work". This means that risk responses aren't implemented, nothing happens to tackle the real risks faced by our project, and risk remains unmanaged.

In recent years, this problem has been recognised by risk management practitioners and project management professional bodies alike. Consequently, a new step has started to appear in the risk processes described in project management bodies of knowledge[*] as well as in risk management standards and guidelines.[†] This makes it an integral part of risk management to make sure that risk responses and actions are implemented, that we do what we said we'd do, and that as a result we minimise our threats, maximise our opportunities, and change our level of overall project risk exposure.

For this reason, there's a strong argument for saying that *risk response implementation is the most important step in the risk management process*, because this is where we actually start to manage the risk! ***The purpose of risk response implementation is to ensure that agreed risk response strategies and actions are implemented effectively.***

If we want effective implementation of our risk responses, there are several important *principles* that we need to follow.

## Don't cut corners

There's a natural tendency for energy and engagement levels to drop off over the course of an extended process. People may start the risk process excited at the prospect of identifying and prioritising the real threats and opportunities in their project. Their level of interest may be slightly reduced when they're asked to develop effective responses for each risk. By the time we get to the implementation phase, however, it's common for people to want to move on to something else, and not spend more time thinking about risk. But this is precisely the time when we need full energy and engagement, as we embark on actions that will actually change our risk exposure, by avoiding or minimising threats, exploiting or enhancing opportunities, and addressing the level of overall project risk.

Having spent considerable effort in the earlier steps of the risk process, it would be a mistake to cut corners at this stage. We need full commitment

---

[*] (Association for Project Management, 2004, 2012; Project Management Institute, 2017)
[†] (Office of Government Commerce, 2010)

to implement agreed risk response actions properly, giving them our complete attention and focused action.

## Motivate action owners

If we want to be sure that agreed risk response actions are implemented well, we need a high level of motivation from our action owners. These are the people who are being asked to take on additional tasks, above their routine project activities, in order to manage risks that may or may not occur in the future. If risk owners simply drop risk actions on action owners without properly explaining why this is important, they shouldn't be surprised if action owners are less than fully enthusiastic about agreeing to do additional work.

The next section suggests some ways in which risk owners might motivate action owners.

## Assume nothing

Risk owners might be forgiven for assuming that action owners will do what they agreed to do, allowing risk owners to pay no further attention to the matter. But implementation of agreed risk response actions is so important that we should assume nothing and check everything. Of course, a balance must be struck between proper delegation and over-zealous micromanagement, but it may not be wise to simply pass on details of risk actions and hope that action owners will pick them up and implement them fully.

## Summary

The purpose and principles of risk response implementation are summarised in Table 9-1.

Table 9-1 Purpose and Principles of Risk Response Implementation

| Purpose | To ensure that agreed risk response strategies and actions are implemented effectively |
|---|---|
| Principles | • Don't cut corners<br>• Motivate action owners<br>• Assume nothing |

## JUST DO IT!

In many ways this step in the risk process is the simplest, but it is also the most difficult. It is simple because it just involves doing what we said we would do in the risk response planning step. If we chose an appropriate risk response strategy for each individual risk and for overall project risk exposure; if we assigned the right risk owner to be responsible for managing the risks; and if they translated the chosen strategy into specific, well-defined risk actions, each with an action owner, clear completion criteria, an allocated budget, timeline, and resources—then all the pieces are in place for our individual threats and opportunities to be managed appropriately and the overall level of project risk to be addressed. We only have to put it all into practice. Simple, right?

The difficulties begin when we ask project team members and others to take action in order to manage individual risks or overall project risk exposure. Common responses include:

- "I'm far too busy doing my regular project work to spend time on addressing risks."
- "Risk management isn't my job."
- "Risk actions are extra work, outside our agreed scope."
- "Most risks probably won't happen, so why should I waste my time on them?"
- "Let's wait and see if any of these risks occur, and deal with them if they do."
- "Is this the best use of limited project funds or time or resources?"

There's a well-known story about four people named Everybody, Somebody, Anybody, and Nobody:

> *There was an important job to be done and Everybody was sure that Somebody would do it.*
> *Anybody could have done it, but Nobody did it.*
> *Somebody got angry that Nobody did it, because it was Everybody's job.*
> *Everybody thought Anybody could do it, but Nobody realised that Everybody wouldn't do it.*
> *In the end, Everybody blamed Somebody when Nobody did what Anybody could have done.*

If we are to avoid this happening with implementation of agreed risk response actions, then the risk owner needs to exercise their powers of persuasion and leadership, motivating their action owners to implement agreed risk actions. There are several ways to achieve this:

- *Involve action owners in developing actions.* Part of the role of risk owner is to ensure that action owners understand what they are being asked to do and why, and that they accept this responsibility. It can be helpful if risk owners and action owners develop risk actions together, involving the action owner in specifying what is required. If action owners have had a hand in deciding what they'll be doing, it's likely to lead to higher levels of engagement.
- *Explain the benefits of action.* Both carrots and sticks can motivate, if used properly. The "carrot" side of the equation here involves ensuring that action owners are fully aware of why they are being asked to take on these additional tasks. The specific risk actions that they will implement will protect the project from the effect of unmanaged threats, allow us to turn opportunities into real savings for the project, and make everyone's life easier by reducing the level of overall project risk exposure. This message will be more powerful if it presents the action owner with specific benefits that will arise if their risk actions are successfully completed.
- *Explain the consequences of non-action.* Conversely, the "stick" message makes the results of non-action clear to the action owner. If we don't manage this threat, we could have this problem, with additional cost, delay, or hassle. If we miss this opportunity, we won't get the potential saving of time, effort, or money. Explain why this risk response action matters, and the action owner will know why they need to pay attention to implementing it fully.
- *Provide all necessary resources.* Unfortunately, action owners are often asked to take on additional tasks to address risks, without being given any extra time, effort, or resource to do the extra work. This is bound to be demotivating! Risk response actions will have been fully specified during the risk response planning stage, including the duration, budget, and resource requirements. When the risk action is passed to the action owner, the risk owner must also make sure that the necessary resources go with it.
- *Demonstrate and role-model required behaviours.* Often, risk owners are also action owners, when they are the person best placed to manage a particular risk. This means that risk owners can show what a good action owner looks like, which will be helpful when they come to ask others to become action owners. When an action owner can see that the risk owner takes this seriously and is personally committed to effective implementation of risk actions, they are more likely to behave in the same way.
- *Celebrate success.* When an action owner successfully completes an agreed risk response action, it can result in removing or reducing an individual threat, or in capturing or enhancing an individual opportunity. It might also lead to a significant improvement in overall project risk exposure. In any of these cases, the risk owner should celebrate that success. Name the

action owner in the next risk report, tell others what a good job they did, thank them personally—these are all highly motivating and will make the action owner want to do it again, as well as encouraging other action owners to perform well.

Each of these motivational approaches applies equally well to ensuring the implementation of risk actions targeting threats or opportunities or overall project risk. In many ways, the motivation challenge for the risk owner is easier when encouraging action owners to address threats, because there is a natural inclination towards self-preservation and self-protection. This means that the risk owner needs to work extra hard to ensure that action owners spend time and energy on opportunities, which can be seen as "optional extras".

In both cases, the key point is that risk actions are not extra work, additional to "real project work". Risk actions **are** real project work—they are tasks that need to be completed in order to give our project the best possible chance to succeed in meeting its objectives and delivering its benefits. They should therefore be treated like any other project task. Each risk response will have been fully defined during the previous step of the risk process, with a duration, budget, resource requirement, completion criteria, etc. The defined action will have been allocated to an action owner who has the necessary ability and availability to complete it. A new task should have been added to the project plan for each agreed risk response, and these should be completed, reviewed, and reported on like all other project tasks.

An important part of this risk response implementation step is to monitor the effect of actions after they have been taken. For example, the "risk-effectiveness" of each proposed response should have been considered during risk response planning, as an indication of the change in risk exposure which can be expected as a result of implementing the chosen response. Having completed the action, the risk owner and/or action owner should assess the actual result to decide whether the risk has been changed in the manner predicted. The status of actions and their results are documented in the risk register. Where planned actions have not been fully effective as expected, the risk owner may decide that additional actions are needed, and these should also be recorded in the risk register.

One possible side-effect of taking action to address risks is particularly important. In some cases, actions taken in response to one risk may introduce new risks that previously did not exist. Such risks are called "secondary risks", not because they are less important, but because their existence is dependent on a prior action being completed. Some secondary risks will have been identified during the risk response planning step, and this is discussed in detail in Chapter 8. But more secondary risks may come to light when responses are actually implemented, and these should be identified, documented, and addressed as part of the risk review step (see Chapter 11).

## WRITE IT DOWN

A number of things change as a result of implementing agreed risk response actions, and these need to be recorded in the risk register. Most important is to document the status of risk actions, to identify any additional actions that may be required, and to list any identified secondary risks that have become active as a result of taking action. There are already fields in the risk register covering actions and secondary risks, and these fields need to be updated with the current status, as shown in Table 9-2.

The risk owner is responsible for updating the risk register fields for those individual threats and opportunities that they own, taking input from action owners as necessary. The project manager is responsible for ensuring that risk actions aimed at managing overall project risk are implemented, and progress can be recorded in an update to the risk report.

Table 9-2 Risk Data to Record after Risk Response Implementation

| Data Field | Risk Register Level | | |
| --- | --- | --- | --- |
| | Minimum | Typical | Detailed |
| Risk Response Data [Updated] | | | |
| Summary of risk actions:<br>• For existing actions:<br>   ○ Action status [UPDATED]<br>• For new actions:<br>   ○ Action owner<br>   ○ Task description | X | X | |
| Detailed risk actions:<br>• For existing actions:<br>   ○ Action status [UPDATED]<br>• For new actions:<br>   ○ Action owner<br>   ○ Task description<br>   ○ Completion criteria/deliverables<br>   ○ Action window<br>   ○ Duration/budget/resource requirement | | | X |
| Secondary risks (if any) [UPDATED] | | | X |

## SUMMARY AND REFLECTION QUESTIONS

Although it's clearly vital that agreed risk responses should be implemented, this is often not explicitly included in the risk process as followed by many organisations. Instead, an assumption is made that actions once agreed will of course be implemented. This is dangerous, because the project might proceed on the basis that its risks are being managed effectively, when in reality the risk exposure remains unchanged. Without taking effective action, the project will still be at risk.

A well-known riddle* illustrates this problem well:

*There are five frogs sitting on a log, and four decide to jump off.*
*How many frogs are on the log?*

The obvious (and wrong) answer is one. A less obvious (but equally wrong) answer is none (the four jumping frogs disturb the log and the fifth one falls off). The right answer is five. The frogs just decided to jump but they didn't. Deciding is not the same as doing.

All the previous steps in the risk process lead up to deciding what to do, as described in Chapter 8. But unless we take action, our situation remains unchanged and we are just as exposed to risk as we were at the start of the risk process. If we don't jump, we remain where we were. We have to get the frogs off the log.

Of course, jumping off the log may not solve all our problems or deliver all the expected benefits. We may find new threats or opportunities after we've jumped (crocodiles in the water, or a new food source), and these will require identification, analysis, and action. Implementing agreed risk responses should deal with the risks we knew about, but residual risks will remain if our actions weren't as effective as we expected, and secondary risks may arise as a direct result of our chosen actions. Our changed situation needs to be reviewed, so that we can discover where we might require further attention and action. Chapter 11 considers how we can determine whether we've done enough by implementing the agreed risk actions. But first we should inform our key stakeholders about what we've done so far to manage the risk in our project. We need to tell others, and this is covered in the next chapter.

## Reflection questions

- *"Risk response implementation is the most important step in the risk management process."* Do you think this is true or not? Explain your reasons.
- Explain the different roles of the risk owner and action owner. Why do we need both roles? How do we decide who is the right person for each role?
- How can a risk owner ensure that action owners do what needs doing?

---

* (Feldman & Spratt, 1998)

## THE NEXT STEP ("NOW WE'VE IMPLEMENTED OUR OPPORTUNITY RESPONSES, WHO SHOULD WE TELL AND HOW?")

A lot of the risk process so far has involved small groups or individuals, identifying risks in workshops or interviews, working together to prioritise individual risks and perhaps also model their overall effect on the project, selecting appropriate risk response strategies, and nominating risk owners who will work with action owners to define and implement specific risk response actions.

As a result, some people in the project have a fairly good idea of the most important threats and opportunities facing the project, as well as the overall level of project risk exposure. They also understand how we plan to tackle individual risks and overall project risk, and the current status of planned risk actions.

But other people have a legitimate interest in the results of the risk process, not just those who have been actively involved in its various stages. We need to ensure that other project stakeholders are informed about what we've found and what we're doing about risk. Everyone knows that communication is important, so we should communicate about risk, and tell others.

# Chapter 10
# Telling Others

We've finally reached the point in the risk management process at which risk has been managed! After identifying and prioritising individual threats and opportunities, and evaluating overall project risk exposure, we've developed and implemented appropriate risk responses, resulting in a change to the level of risk faced by the project. Although this is an important outcome in itself, there are other people who need to know what has been achieved through our risk management efforts. We need to **tell others** about what risks we've found, how risky the overall project is, what actions we've taken, and what level of risk remains.

## PURPOSE AND PRINCIPLES OF RISK COMMUNICATION AND REPORTING

There are many different individuals and groups of people outside the immediate project team who are affected by how risky the project is, or by specific threats and opportunities within the project. These people are known as stakeholders,[*] defined as, *"an individual, group, or organisation that may affect, be affected by, or perceive itself to be affected by a decision, activity, or outcome."*[†]

Not every stakeholder in our project needs the same information about risk, so we need a tailored approach that provides the right information to the right

---

[*] (Bourne, 2009, 2015)
[†] (Project Management Institute, 2017)

person or group, at the right time, in the right format, for the right purpose.[*] There are two reasons for telling others about risk, and they are the same for all stakeholders:

1. *For attention and information.* We need to tell people what they need to know about risk that they don't currently know.
2. *For action and involvement.* We need to advise people about what they need to do about risk that they're not currently doing.

***The purpose of risk communication and reporting is to provide project stakeholders with timely and accurate risk information to support appropriate risk-informed decision-making and action.***

The following ***principles*** are important to remember if we want to communicate effectively about risk.

## Be honest

Honesty is not the *best* policy when it comes to risk communication; it is the *only* policy. We can't just give the "right answer" or tell people what they want or expect to hear. People can only make good risk-informed decisions if they have reliable and accurate risk information. This partly depends on the quality of the analysis, but it also relies on the integrity of the people compiling risk reports.

Often it appears difficult to "speak truth to power", especially when we are communicating about risk and uncertainty. The future is not fixed, and there is no guarantee that the risks highlighted in our risk reports will actually happen. But in fact, this is the whole point of risk communication and reporting—to expose and explain the degree of uncertainty associated with our project and the activities within it. We can't be expected to provide absolute certainty when we are talking about uncertainty!

The problem seems more acute when risk reports are perceived as bringing "bad news", warning of adverse events and circumstances that could make it hard to achieve our project objectives, but which may never happen. It's true that communicating this type of message can be challenging, but consider the alternative. If the risk process has uncovered an unacceptable level of overall project risk exposure, or found one or more showstopper threats, then it is irresponsible to keep quiet. The professionalism of the project manager and risk specialist demand honesty. By being prepared to communicate openly with

---

[*] (Bourne & Weaver, 2016; Hillson, 2011)

project stakeholders about unwelcome topics, we give them the time and space needed to consider their options and take appropriate action.

This is a case where the inclusion of opportunities in the risk process makes a real difference. A threat-only risk process can only bring bad news. By contrast, when our risk reports also contain lists of potential events and circumstances that could save time and/or money, enhance performance or reputation, and make it easier to achieve our project goals, then these reports will be welcomed with open arms by our project stakeholders. Some project managers prefer to keep quiet about the opportunities in their project as well, thinking that they can keep them in reserve "for a rainy day". If something bad happens or we get behind schedule or over budget, then perhaps we can work on one or more of these hidden opportunities and counteract the downside impact of threats. But this too is unprofessional. By not reporting on opportunities, we deny project stakeholders and decision-makers the information that would enable them to make fully risk-informed decisions, balancing both upside and downside risks.

It is much better to be completely honest about the risk exposure as it currently stands on our project, including the worst threats and the best opportunities. Then and only then can the risk management process fulfil its purpose of allowing risk to be managed proactively and effectively.

## Be specific

When it comes to risk reporting, one size does not fit all. Different stakeholders need different risk information to help them make appropriate risk-informed decisions within their area of interest. Too often, the output from the risk process consists of a risk register and sometimes an accompanying risk report, which are distributed to all stakeholders regardless of whether the content meets their needs. Even worse is when the risk register and report aren't distributed but merely made available on a project intranet or shared drive, with the intention that anyone who needs to know anything about risk on this project only has to go and look for themselves.

Tailoring of risk outputs for specific stakeholders is essential. Some need lots of detail while others prefer summaries. Some need to be notified about risks instantly, but others can wait for a routine progress update. A written report suits some stakeholders, some like a face-to-face briefing or presentation, while a dashboard format works for others. For some of our stakeholders, only a subset of risks is relevant (contract risks, technical risks, financial risks, etc.), but others need to see the whole picture. This means that we can't rely on the risk register and a standard risk report to meet the information requirements of all our stakeholders. We need to produce a range of specific outputs, each tailored to communicate effectively with one or more stakeholders.

Of course, this takes more work and effort! But we've worked very hard throughout the risk process to identify, understand, and prioritise risks; to determine what they mean for the whole project; and to develop and implement effective risk responses. All that hard work will be wasted if the risk information we've produced doesn't reach the people who need to know about risk in order to make good decisions, or if the risk information they do see doesn't help them. Effort spent on producing specific tailored risk outputs is an essential investment in ensuring that risk information is used, and that it doesn't lie hidden in an unopened risk register or an unread risk report.

## Be timely

Late information is as bad as wrong information or no information at all. There's no point in passing on information about a risk to a decision-maker with a footnote saying, "*We discovered this risk last month, but we've only just got around to telling you about it. Sadly, the chance to influence the risk ended last week. If you'd known about it sooner you might have done something, but . . .*".

Interestingly, risk information provided too early can also be unhelpful. If we uncover a risk that might affect our project in a year's time, and we can't take meaningful action to manage it until a week or two before it's due to occur, telling the potential action owner a year in advance probably isn't useful. They're likely to forget, or move on to another role, well before the risk gets close.

It's important to communicate about risk at the right time, ensuring that project stakeholders get timely information allowing them to make effective decisions. We should report in "Goldilocks Time": not too early, not too late, but just right!

## Summary

The purpose and principles of risk communication and reporting are summarised in Table 10-1.

**Table 10-1 Purpose and Principles of Risk Communication and Reporting**

| Purpose | To provide project stakeholders with timely and accurate risk information to support appropriate risk-informed decision-making and action |
|---|---|
| Principles | • Be honest<br>• Be specific<br>• Be timely |

## STAKEHOLDER RISK INFORMATION NEEDS ANALYSIS

Each stakeholder has a different requirement for risk-related information, and the risk process should recognise this and deliver timely and accurate information at an appropriate level of detail to support the needs of each stakeholder. It is not adequate to simply take raw risk data and pass them on, or to make the same standard risk information available to all project stakeholders.

In order to communicate appropriately about risks to our project stakeholders, we need a structured approach. The key is to identify those stakeholders who need risk information, and to define their information needs. Each stakeholder has a specific interest or "stake" in the project, and this will affect what risk information they need to receive. We can then design communications to meet those needs specifically and precisely, using different outputs from the risk process.

Stakeholders should have been identified during the project initiation step, but if that wasn't done, then we need to do it here. Most projects have a common set of stakeholders, as shown in the left-hand column of Table 10-2. In addition, other types of stakeholder might be present in specific project types, listed on the right-hand side of Table 10-2.

Table 10-2 Common and Specific Project Stakeholders

| Project Stakeholders | |
|---|---|
| Common to Most Projects | Specific to Some Projects |
| Project sponsor/business owner | Suppliers/subcontractors |
| Project manager | End users |
| Project team | Regulators |
| Customers/clients | General public |
| Senior management | Competitors |
| Program/portfolio manager | Pressure groups/lobbyists |
| Other projects/operations | Politicians |

Once we've listed our stakeholders, we can analyse the risk information needs of each individual or group, linked to their interest or "stake", by answering the following questions:

- What risk information is required?
- How will it be used?
- What level of detail and precision is required?
- When must risk information be supplied?
- What time delay is acceptable (if any)?
- How frequently are updates needed?
- How should information be delivered?

**Table 10-3 Stakeholder Risk Information Needs Analysis, with Sample Entry**

| Stakeholder | Interest ("Stake") | Information Required | Purpose | Frequency | Format |
|---|---|---|---|---|---|
| Project Sponsor | Meet business case | Project status summary | Project monitoring | Monthly | Two-page hard-copy report |
| | | Risk status summary | Manage key risks | Monthly | Two-page hard-copy report |
| | | Problems outside project control | Assist project manager | Immediate | Verbal plus email |
| | | | | | |
| | | | | | |

This information should be recorded in a stakeholder risk information needs analysis, such as the example shown in Table 10-3.

## RISK COMMUNICATION DESIGN

Based on the stakeholder risk information needs analysis, we can design tailored risk communications that meet the identified requirements of each stakeholder individual or group. The best way to do this is to take the various standard outputs that we've generated during the risk process so far, and combine these in various ways to create something specific for each stakeholder. It's even better to design reports in a hierarchical manner if possible, with high-level reports compiled from summaries of low-level reports, in order to reduce the overhead of producing multiple outputs.

Standard outputs from the risk process include the following:

- For individual threats and opportunities
  - Risk register
  - "Top risks" list
  - Individual and aggregated risks for escalation
  - Risk distributions
  - Metrics and trend analysis
- For overall project risk (from quantitative risk analysis)
  - Risk report

These outputs are described in more detail later in this chapter.

We can combine these various outputs into an information package for delivery to stakeholders, most commonly as some form of risk report. The level of detail needed by different stakeholders will of course vary, and three levels of risk report may be considered:

- *Risk List.* This is the minimal level at which risk information can be reported, presenting nothing more than a simple list of identified risks, perhaps prioritised by probability and impact, and maybe filtered to show only those risks currently active. Threats and opportunities may be listed separately or combined into a single list. An example format for a risk list is shown in Figure 10-1.
- *Summary Risk Report.* At this level the risk report is likely to contain only basic information. An executive summary should be used at the start of the report, followed by a summary of the level of overall project risk exposure in a few concise paragraphs. This should be followed by details of the key risks, describing their main features and the planned responses. The remaining risks may not be discussed in detail, but could be presented using graphical analyses and pictures. Key changes since the last report should be summarised. The Summary Risk Report should end with conclusions and recommendations, briefly assessing the current situation and summarising those actions which are required to keep risk exposure within acceptable limits. The risk register or "Top risks" list may be supplied as an appendix to the Summary Risk Report. Figure 10-2 provides an example contents list for a Summary Risk Report.
- *Detailed Risk Report.* This type of report will contain full details and analysis of all identified risks, with supporting information. The report should start with a brief executive summary, then summarise the level of overall project risk exposure. A review of current risks may follow, highlighting the top threats and opportunities. A detailed analysis of individual risks may be presented, covering changes in the status of identified risks, and listing any new risks identified since the last report. Risk distribution data can be presented and discussed, identifying common sources of risk, hot-spots of exposure in the project, etc. Quantitative analytical results may also be presented and discussed where quantitative risk analysis has been undertaken. Key risk themes will be identified and discussed, as well as a trend analysis addressing changes in risk exposure. Similar to the Summary Risk Report, the more detailed report will also end with conclusions (what does it mean) and recommendations (what should be done), though these are likely to be more detailed in the Detailed Risk Report. Supporting information will also be presented in appendices, including the risk register, other detailed assessment results, and supporting data. Figure 10-3 gives an example contents list for a Detailed Risk Report.

Once we've worked out the content of the risk information package that we'll provide to each stakeholder, we need to think about how to deliver it. Part of our initial stakeholder risk information needs analysis is to determine how they prefer risk information to be delivered, and we should bear this in mind

| Project Number: | **PX42651/SC3** | Client: | **MegaCorp Inc** | Review date: **1 April 2019** |
|---|---|---|---|---|
| Project Title: | **ASCENT System Upgrade** | Project Manager: | **Seymour Goodnews** | |

| Risk Ref. **T = threat** **O = opportunity** | Rank | Risk Title and Description | Cause | Impact | Probability (L,M,H) | Impact on Time (L,M,H) | Impact on Cost (L,M,H) | Impact on Performance (L,M,H) | Action(s) | Action Owner |
|---|---|---|---|---|---|---|---|---|---|---|
| T00175 | 1 | **Requirement volatility.** The plan assumes that there are no significant changes in requirements, but the client may request or require change after design freeze. | Client has not yet defined user spec. | Rework to design documentation, or repeat work if later in project. | M | H | L | H | Ensure system is as generic as possible to allow for likely level of changes. | System Design Manager |
| | | | | | | | | | | |
| | | | | | | | | | | |
| | | | | | | | | | | |

**Figure 10-1** Example Risk List Format (Prioritised)

| |
|---|
| Executive Summary |
| Overall Project Risk Status (at this review) |
| Top Risks with agreed actions and owners |
| Key Changes since last review |
| Conclusions & Recommendations |
| Appendix: Risk Register |

Figure 10-2  Summary Risk Report Example Contents List

wherever possible. There are many alternative delivery methods, including written reports in hard copy or electronic format (email, intranet, website, accessible databases, social media postings), verbal reports (briefings, presentations, progress meetings, video-casts), graphical or numerical outputs (tables, charts, posters), etc.

Once we've designed the content of each risk communications deliverable, we also need to nominate a defined owner responsible for its production, and an accountable person who is the designated approval authority. It may also be

| |
|---|
| Executive Summary |
| Overall Project Risk Status (at this review) |
| Top Risks with agreed actions and owners |
| Risk Distributions<br>• High/Medium/Low risks<br>• Causal analysis (mapped to RBS)<br>• Effects analysis (mapped to WBS) |
| Quantitative Risk Analysis [where used]<br>• Model inputs and structure<br>• Key outputs and analysis |
| Changes since last review, including metrics and trend analysis |
| Conclusions & Recommendations |
| Appendices<br>• Risk Register<br>• Quantitative risk analysis data [where used]<br>• Other detailed supporting data |

Figure 10-3  Detailed Risk Report Example Contents List

helpful to identify those who we'll need to consult for their input, and those who will receive the output for information. A RACI (Responsible, Accountable, Consulted, Informed) analysis might be used to document this.

Proposed risk communications should be documented as shown in Table 10-4, and this can either form part of the project's Communication Plan or be included in the Risk Management Plan. We can also cross-check Table 10-4 (risk outputs) against Table 10-3 (stakeholder requirements) to ensure that all stakeholder requirements are met by the proposed outputs.

**Table 10-4 Risk Outputs Definition, with Sample Entry**

| Title | Content | Distribution | Frequency | Format | Responsibility |
|---|---|---|---|---|---|
| Risk status summary | Summary of risk status<br>Key risks<br>Changes since last report<br>Recommended actions | Project Sponsor<br>Project Manager<br>Programme Review Board<br>Corporate Risk Manager | Monthly | Two-page hard-copy text report, including "Top risks" table, risk trend graph, and Action table. | Project Manager (with input from key team members) |
|  |  |  |  |  |  |
|  |  |  |  |  |  |

## NOW COMMUNICATE!

After we've identified the risk information needs of each stakeholder and designed tailored risk communications for each one, we need to produce the set of outputs and deliver them to stakeholders. After one or two cycles of risk reporting, the risk communications process should be reviewed with key stakeholders to check whether their needs are being met, or whether adjustments are required. Periodic reviews of the risk communications process should also be planned, because the risk information needs of stakeholders are likely to change with time.

## DELIVERY TIMELINESS

One of the key issues in communicating about risk is to identify when stakeholders require information to be supplied, and timeliness is as important as content. There's no advantage in getting the right information too late to use it or too early to be relevant. It's therefore important to consider when risk reporting should be undertaken in the risk process. Risk methodologies tend to include risk reporting as one of the last steps in the process. In reality, however, risk information becomes available incrementally throughout the risk process.

For example, it is possible to list the risks immediately after risk identification, and to start compiling the risk register using output from the qualitative risk assessment. A preliminary risk report could therefore be produced early in the risk process, without waiting for all the details, and then reissued when more information becomes available.

Given the need to inform stakeholders promptly about risks requiring their attention or action, we could consider producing risk reports as soon as possible, even if some of the information is preliminary and will require updating later. Early reporting maximises the time available for consideration of options, development of effective responses, and implementation of planned actions. Any delay in communicating risk information reduces the available time in which it is possible for stakeholders to respond proactively to risk.

## COMMUNICATING ABOUT OPPORTUNITIES: TOGETHER OR SEPARATE?

There's one key question to answer when designing risk communications from a process that includes opportunities alongside threats: *Should we report them together or separately?* The simplest answer might be to assume that we can use a common report format for both upside and downside risks, but this is not always helpful. In this section, we'll consider when it's useful to produce a single output for both types of risk, and when we need to keep them apart. We can think about this for each of the main types of output from the risk process, including the following:

- Risk register
- "Top risks" list
- Risk distributions
- Metrics and trend analysis

Many of these outputs will have been produced during the qualitative risk assessment step of the process (see Chapter 6), and these can be easily and simply incorporated as components of our risk communications.

### Risk register

The risk register is a document or database holding information on all identified risks in a common format. The level of detail can vary depending on the complexity of the project and the depth of the risk management process. A simple risk register might record only a single line for each identified risk, including, for example, a risk identifier code, a descriptive title for the risk, assessments of probability and impact, the preferred response strategy, and the risk owner responsible. These simple formats are often created in standard office packages such as Microsoft® Excel®. More complex risk register formats present detailed information on all risks, and there are many proprietary software tools that support risk register production. We've seen in earlier chapters how the data held in a risk register are built up as the risk process progresses, and how different levels of detail are possible in the risk register.

For a risk process that includes both threats and opportunities, the obvious question is whether we should hold data on both in a single risk register, or have two registers.

In principle, because the key characteristics of all risks are the same whether they are threats or opportunities, they can both be described within a single risk register. If we do this, it's a good idea to use a numbering scheme which indicates the risk type, perhaps using a prefix of T for threat and O for opportunity (as in Figure 10-1, for example).

The same data elements in a risk register apply to both threats and opportunities, although there are some important differences. For example, descriptions of impacts will be negative for a threat (delay, extra cost, performance shortfall, etc.), whereas impacts for an opportunity will be positive (saving time or cost, enhancing performance, etc.). Different response strategies also apply to threats and opportunities (as described in Chapter 8).

On balance, it's better to maintain a single risk register containing data on both threats and opportunities, then to apply filters to extract data subsets as required. Apart from anything else, the overhead on maintaining risk register data is lower if we only have one of them!

## "Top risks" list

This is a prioritised risk list showing the worst threats and best opportunities, usually ranked by Probability-Impact (P-I) Score. Many organisations present the "Top 10", although it may be unwise to limit the number of "top" risks artificially to 10—the number on the list should reflect those risks requiring urgent or focused risk management attention. Projects are often surprised by the risks ranking just outside the "Top 10", which could still have a significant effect, but which were not receiving the same attention as the "Top 10".

In the same way as for the risk register, the question is whether to have a combined "top risks" list, or to present "top threats" and "top opportunities" separately. The "top risks" list is used to focus attention and action on the most significant and important risks, so it makes sense to combine both the worst threats and the best opportunities into a single list. It may be the case that the top three or four risks are all opportunities, which rank higher than the highest-ranking threat, so having a combined "top risks" list ensures that we concentrate on the most important risks, regardless of whether they are threats or opportunities.

## Risk distributions

Useful management information can be generated from considering the distribution of identified risks across various categories, as analysed during the qualitative risk assessment phase (discussed in Chapter 6). Three particular types of risk distribution can give us useful information: risk priority, project breakdown structures, or time windows. These are outlined below:

- *Risk priority.* The Probability-Impact Matrix (P-I Matrix) (see Figure 4-5) plots risks by probability/frequency and impact, and divides them into priority categories (Red/Yellow/Green or High/Medium/Low priority). Risk reporting outputs might include the numbers of risks in each category, or perhaps the total P-I Score for the risks in each category. The distribution of risks across these categories gives an indication of the overall level of risk faced by the project.

    We can also track trends in risk exposure by monitoring the movement of risks between priority categories. The aim is to move threats from Red/High to Green/Low, and to improve Green/Low opportunities towards the Red/High zone. We can plot changes in the number of risks in each priority category over time, as shown in Figure 10-4, indicating changes in numbers of threats and opportunities between one risk assessment and the next. The preferred distribution is High opportunity/Low threat. Figure 10-4 shows that the number of opportunities has increased between Issue 1 and Issue 2 of the risk report, and there are more high-priority opportunities as well. We can also see that the number of threats has reduced for Issue 2, and the threat distribution has changed with mostly low-priority risks.
- *Project breakdown structures.* Chapter 6 described how identified risks can be mapped against the various breakdown structures in the project, to reveal areas in which risk exposure is concentrated. Each breakdown structure mapping reveals something different about the distribution of risk across the project:
    - Risk Breakdown Structure (RBS)—this allows root-cause-analysis, revealing common causes of risk. We can then identify RBS elements

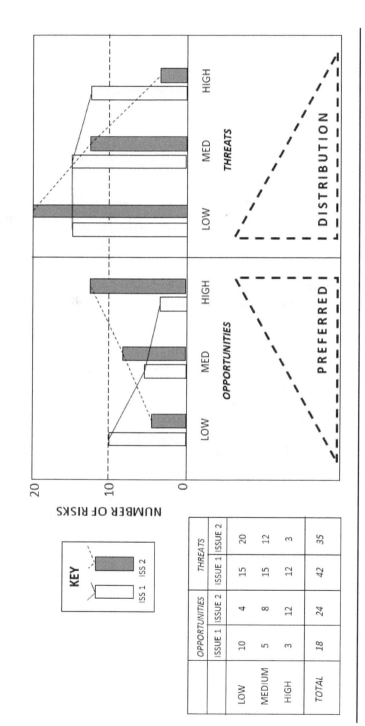

**Figure 10-4** Risk Distribution by Priority Category

that are either major sources of threat or significant sources of opportunity.
- ○ Work Breakdown Structure (WBS)—this can expose hot spots of risk on the project, which are those parts of the project with the highest level of associated risk. We may decide to concentrate our best resources on these areas, or include additional float and contingency, or adopt a different development approach.
- ○ Cost Breakdown Structure (CBS)—identifying which line items are most risky will help us to allocate contingency funds or take other measures to protect the budget.
- ○ Organisational Breakdown Structure (OBS)—understanding who is responsible for risky areas of the project will help us to assign risk owners. It might also tell us if particular team members may need assistance or support, either because they have a significant risk exposure in their part of the project, or because additional risk actions might distract them from other important project work.
- ○ Product Breakdown Structure (PBS)—knowing where variability in project deliverables could occur will guide us in design, development, testing, and audit activities.
- ○ Benefits Breakdown Structure (BBS)—identifying possible deviations in project outcomes will help us to manage stakeholder expectations.

In each of these cases, it's a good idea to separate threats and opportunities, allowing us to map the distribution of both negative and positive risk exposure.
- *Time windows.* A useful element of risk communication is to report on "impact windows" (when the risk might affect the project if it were to occur), and "action windows" (when action is possible and likely to be effective). Both of these are described in Chapter 6. Clearly there should be no overlap between the two windows for a given risk, as this would suggest that actions might not be completed before the risk could occur. Plotting both impact and action windows on the same chart would show if this situation could arise, as illustrated in Figure 6-7. It would also indicate periods in the project when risk response activity might be highest, which might have implications for project resourcing. Finally, such an output would show the time periods when the project is most at risk from the potential effects of uncertainty (either positive or negative), assisting decisions in allocation of contingency to the project (cost, time, resource, performance, etc.), and perhaps indicating the need to consider scheduling strategies such as parallelism or crashing.

Impact window/Action window analysis doesn't need to distinguish between threats and opportunities, and it is common to use a single plot for both types of risk.

## Metrics and trend analysis

We can use various metrics to monitor changes in risk exposure over time,[*] including:

- Number of active risks ($N_{ACTIVE}$)
- Number of closed risks ($N_{CLOSED}$)
- Total P-I Score for active risks (T)
- Average P-I Score for active risks ($A = T / N_{ACTIVE}$)

Plotting changes in any of these four metrics over time is simple enough, and this clearly needs to be done separately for threats and opportunities.

In addition to these four simple metrics, we can combine T and A into a composite measure known as the Relative Risk Exposure Index (RRE). Risk exposure is more than just the total number of the risks which face the project ($N_{ACTIVE}$); it also needs to take account of their importance (indicated by both T and A). A project might be more at risk with twenty small risks than with three large ones. RRE accounts for both number and severity, and is calculated by taking the product of Total P-I Score (T) and Average P-I Score (A) at the current point in time ($T_c * A_c$) and dividing it by a baseline value from the start of the project or phase ($T_b * A_b$):

$$RRE = \frac{(T_c * A_c)}{(T_b * A_b)}$$

The resulting value of RRE is 1.0 if the current level of risk exposure is the same as the baseline, with <1 indicating reduced risk exposure compared to the baseline, and >1 meaning increased risk exposure over the baseline.

RRE calculation only works if all the risks included in the calculation are of the same type (either all threats or all opportunities), so two metrics are usually produced. Trends in these two versions of RRE also mean different things. Higher values of $RRE_{THREAT}$ mean worsening exposure to threats compared to the baseline situation. Changes in $RRE_{OPP}$ mean the opposite, since higher values mean increased opportunity.

Having defined various metrics for risk exposure and measured them as the project proceeds, we can then perform trend analysis and report on the way risk exposure is changing on the project.

Examples of risk metrics and trends are given in Figure 10-5, with separate figures for threats and opportunities. The data in this table are presented graphically in Figure 10-6. This shows that the trend in overall threat exposure

---

[*] (Hillson, 2004)

| Risk Register Issue | Threat Data ||||| Opportunity Data |||||
|---|---|---|---|---|---|---|---|---|---|---|
| | Number of active threats ($N_t$) | Number of closed threats | Total Threat P-I Score ($T_t$) | Average Threat P-I Score ($A_t = T_t/N_t$) | Relative Risk Exp Index $RRE_{THREAT}$ | Number of active opps ($N_o$) | Number of closed opportunities | Total Opp P-I Score ($T_o$) | Average Opp P-I Score ($A_o = T_o/N_o$) | Relative Risk Exp Index $RRE_{OPP}$ |
| Issue 1 | 20 | 0 | 15.5 | 0.8 | 1.00 | 5 | 0 | 2.3 | 0.5 | 1.00 |
| Issue 2 | 22 | 5 | 18.2 | 0.8 | 1.25 | 8 | 0 | 4.8 | 0.6 | 2.72 |
| Issue 3 | 30 | 12 | 21.7 | 0.7 | 1.31 | 9 | 2 | 7.5 | 0.8 | 5.91 |
| Issue 4 | 23 | 23 | 16.1 | 0.7 | 0.94 | 13 | 4 | 6.2 | 0.5 | 2.79 |
| Issue 5 | 14 | 35 | 9.3 | 0.7 | 0.51 | 6 | 11 | 5.0 | 0.8 | 3.94 |

**Figure 10-5** Example Risk Metrics Data

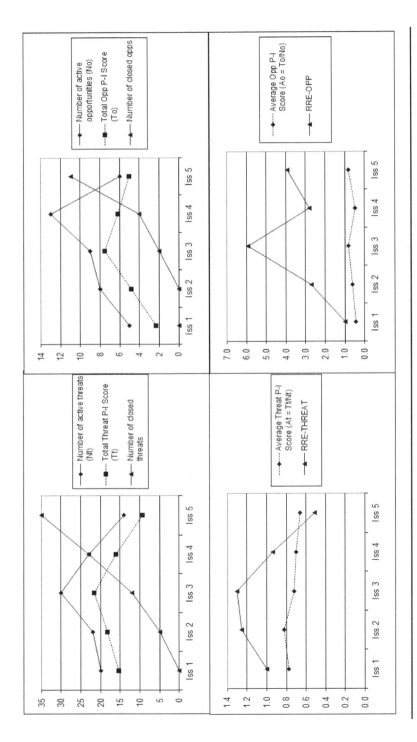

**Figure 10-6** Risk Metrics and Trends (Data from Figure 10-5)

($RRE_{THREAT}$) follows the change in number of active threats since the average P-I Score remains nearly constant, whereas the shape is different for opportunity exposure ($RRE_{OPP}$). The main difference is between Issue 3 and Issue 4 of the risk assessment, when threat exposure is seen to reduce from >1 to <1 (a good thing), but the opportunity index halves from nearly 6 to <3 (a bad thing). This result may suggest a need to refocus the risk process on opportunity management. The closing position at Issue 5 shows threat exposure about half that at Issue 1, while opportunity exposure is about four times higher.

## SUMMARY AND REFLECTION QUESTIONS

A risk process without risk communication and reporting is like a hi-fi without speakers, an MP3 player without headphones, or a computer without a screen. The most amazing analysis can have taken place, but without the means to make it available to others, it is useless and wasted.

Understanding and managing risk is important to project success, including individual threats and opportunities, as well as the level of overall project risk exposure. The risk process uncovers information that we need to know but that we don't currently know, and it suggests risk response actions that we need to do but that we're currently not doing. The purpose of risk communication and reporting is to enable project stakeholders to *pay attention and take action*. This is too important for us to treat lightly. Having spent a lot of time and effort in the earlier phases of the risk process, we need to make sure that we communicate risk information effectively.

### Reflection questions

- Why shouldn't we simply expect anyone who needs information about risk in our project to look in the risk register or read a generic risk report?
- When is it useful to combine information about threats and opportunities in a single report format, and when should they be reported separately?
- What are the dangers in providing risk information to stakeholders at the wrong level of detail, in the wrong format, or at the wrong time?

## THE NEXT STEP ("NOW WE'VE TOLD PEOPLE ABOUT CURRENT RISKS, HOW DO WE KEEP IT UP TO DATE?")

Phew! We've finally completed a full cycle of the risk process. We started by deciding how we were going to approach risk management on our project

(Chapter 4), then we set about identifying individual threats and opportunities (Chapter 5). Next, we prioritised and characterised those individual risks (Chapter 6), analysed their combined effect on overall project risk exposure (Chapter 7), decided what to do about each threat and opportunity and also how to tackle overall project risk (Chapter 8), and took action (Chapter 9). Finally, we designed a set of risk communications and reports to tell others what we'd discovered through the risk process, so that they can take appropriate decisions about risk (this chapter). Are we done now?

Actually, the risk process has two more really important steps, described in the following two chapters. Risk is a dynamic challenge, always changing, coming and going, occurring or fading away. We can't simply run through the risk process once and think we're finished with risk management. As the risk challenge on our project changes, so we need to re-evaluate the level of risk exposure we face, discover what new risks might have arisen, and keep taking action to stay within the risk thresholds for our project. We need to keep things up to date.

# Chapter 11
# Keeping Up to Date

The previous chapters in this section have described one complete pass through the risk management process. We started with deciding the scope and boundaries of risk management on our project (Chapter 4), so that we could adopt an appropriate and effective approach that would address the specific risk challenge we face. As part of the scope, we made an intentional decision to include opportunities in the risk process, alongside threats. Then we identified individual threats and opportunities that might affect our ability to achieve our project objectives (Chapter 5), and prioritised them to find the worst threats and the best opportunities (Chapter 6). After using quantitative techniques to model the effect of individual risks and other sources of uncertainty on our project (Chapter 7), we evaluated the level of overall project risk in terms of probability of success and degree of potential variation in project outcomes. Next, we planned appropriate responses for both individual risks and overall risk exposure (Chapter 8), and implemented those responses in order to actually manage risk (Chapter 9). Finally, we designed and delivered tailored risk communications for our project stakeholders (Chapter 10), giving each stakeholder the specific information that they need so that they can pay attention and take action on risk.

After all this focused effort on risk management, we might be forgiven for thinking that now we've done such a good job, we can move on to some other aspect of managing the project, since the risk challenge has been dealt with. Unfortunately, risks have a habit of re-emerging, resulting in changes to the overall level of risk exposure! And each new risk that arises needs to be identified, analysed and addressed, in order to keep overall project risk within the agreed risk thresholds for the project.

Yes, we may have done a great job on our first pass through the risk process, but we haven't finished yet. As long as the project lasts, the risk challenge remains. We need to **keep up to date** with the risks that we face, stay focused on addressing them appropriately, and not lose momentum.

One final important point before we dive into detail for this chapter. Previous chapters have pointed out where things need to be done differently in order to address opportunities alongside threats, so that we have a fully inclusive risk process. The final couple of steps in the risk process apply equally to threats and opportunities, and there is no real distinction in the way upside and downside risks are handled. So, this chapter and the next both describe what needs to be done for all identified risks, without separate sections for dealing with threats and addressing opportunities—because the approach is the same for both.

## PURPOSE AND PRINCIPLES OF RISK REVIEWS

One of the most important features of risk is that it is dynamic and always changing. As we aim to manage risk effectively on our project, we face a moving target. Nearly all elements of the risk process are time-dependent:

- Risk identification only considers those uncertainties that were faced by the project at a particular point in time, and that we were able to perceive then.
- Identified risks were assessed in the light of the prevailing circumstances at the moment of assessment, and with the information and perceptions that we had in front of us.
- This led us to develop a set of preferred responses which reflected the best option at the time, given what we knew about each risk, what resources were available to tackle it, the status of the project, and so on.

Because risk poses a dynamic challenge, as time passes, so risk changes:

- Some identified threats may no longer exist after a period of time, either because they have been successfully avoided, or because their impact window has passed without the uncertainty actually occurring. Or even worse, the threat might have occurred and become an active problem or issue for the project.
- Similarly, some previously recognised opportunities may not still be there later in the project, either because they have timed out and cannot now

occur, or because they have been successfully exploited and turned into real benefits.
- Risks can get better or worse with time, with higher or lower probability and/or impacts. Either this might be objectively true as circumstances change, meaning that a particular risk really is more or less likely, or it could affect us more or less than we originally thought. Or our perceived assessment of the risk may change as we gain more information or experience on the project. These changes affect both threats and opportunities: A threat might become more or less of a potential problem, and an opportunity could become more or less attractive.
- New risks can emerge which were not previously identified. This may be as a result of shortcomings in the risk identification process which failed to see some of the risks earlier, or it may be because progress on the project and changes in its environment have created new uncertainties that did not previous exist. In some cases, risks may have been present in the project, but were hidden when we looked last time, only emerging after earlier project activities had been completed. We also need to remember secondary risks that we might have introduced when we implemented agreed responses to another risk (sometimes unwittingly, but not always).

The changing nature of risk means that a single risk assessment won't remain valid for the lifetime of our project. We can't allow risk management to be a one-off undertaking; we need to repeat it so that we understand current risk exposure and so that our risk management actions continue to be appropriate and effective. This means that we need some form of **risk review**, either within the risk process, or as part of the project management process.

We need to cover a number of aspects during each risk review, including:

- Tracking changes in the status of identified risks and in the level of overall project risk
- Checking whether agreed risk responses are being implemented as agreed
- Identifying and assessing new risks, and developing appropriate responses for them
- Determining whether the risk process is working effectively

Each of these aspects allows the risk review to fulfil its main purpose, which is ***to provide visibility of current risk exposure***.

As we plan and execute risk reviews in our project, we need to follow a number of important ***principles***.

## Decide when and where to review risk

We could review risk in a separate "risk review meeting", or we might do it as part of the normal ongoing project management process. The approach we adopt depends on the level of risk management applied to our particular project, as defined in the Risk Management Plan (RMP), and both approaches have their advantages. If we hold a separate meeting focused entirely on risk, this reinforces the importance of risk management, imposes some discipline and rigour into the risk process, and gives the project team a specific place to think and talk about how to manage the threats and opportunities associated with their project. On the other hand, including risk on the agenda for regular project progress meetings emphasises that managing risk is an essential part of the way the project is being managed, and it allows the project manager and team to make project decisions in the light of current risk exposure.

Perhaps the right answer is to do both: schedule specific risk review meetings to allow project team members to focus their energy on risk, and also discuss risk at regular project progress meetings, with particular attention to whether agreed risk actions are being completed and what new risks have arisen since last time.

## Review risk regularly

Wherever they are done, risk reviews should be undertaken on a regular basis during the project. The definition of "regular" depends on the level of risk process implemented on the project, which will be defined in the RMP.

As a minimum, risks should be reviewed at key points in the project—for example, at project gateways, major decision points, or transitions between project phases. This ensures that the risks associated with the next phase of the project are identified and understood before proceeding with the project, including those unresolved risks carried forward from the previous phase(s).

Alternatively, risk reviews might be undertaken at regular intervals throughout the project—for example, monthly or quarterly. The frequency should reflect the perceived level of risk on the project, and may vary between different project phases.

## Don't limit reviews to meetings

The problem with reviewing risk at regular intervals is the assumption that nothing urgent happens between risk reviews. But a major new risk might arise at any time, or the status of an existing risk could change unexpectedly. We need to see risk management as a continuous process, and not constrain it to a regular review cycle. The risk process should be flexible enough to allow risks to be identified and addressed whenever they are first perceived, or to develop and implement a different risk response strategy as soon as we realise that it's necessary, rather than waiting for the next risk review meeting or project progress meeting. Nevertheless, regular review points provide a necessary discipline and structure for the risk process—for example, by ensuring that agreed actions are implemented.

## Involve key stakeholders

It's common to involve the main project stakeholders (or their representatives) in the first iteration of the risk process, inviting them to attend a risk workshop to identify and assess individual threats and opportunities, sometimes nominating them as risk owners and involving them in development of risk response strategies, and maybe even asking them to implement risk actions as well. But once the initial risk assessment is completed and we enter into the regular risk review cycle, we tend to forget stakeholders outside the project team, and we don't invite them to contribute to the risk process. This means we lose their input and opinion on changes in current risk exposure, progress in managing previously identified risks, or whether new risks have arisen that they might see but we might miss.

Obviously, it's not practical to invite all project stakeholders to every risk review meeting or project progress meeting as the project proceeds. But we should be aware that they provide an important perspective that can offer us fresh insights into the risks we face, and we should involve them when necessary.

## Summary

The purpose and principles of risk reviews are summarised in Table 11-1.

Table 11-1 Purpose and Principles of Risk Reviews

| Purpose | To provide visibility of current risk exposure |
|---|---|
| Principles | • Decide when and where to review risk<br>• Review risk regularly<br>• Don't limit reviews to meetings<br>• Involve key stakeholders |

## TYPICAL RISK REVIEW TECHNIQUES

A risk review should consider a number of key questions about the risks facing the project, whether it is conducted as a separate meeting or as part of a wider project meeting. It is important to:

- Assess the status of existing risks
- Review the effectiveness of risk responses
- Identify new risks
- Review the effectiveness of the risk process

Each of these steps applies to both threats and opportunities, as described below.

### Assess status of existing risks

One major purpose of the risk review is to consider the current status of each identified individual risk. The typical threat or opportunity has a range of possible status values that it might pass through, as illustrated in Figure 11-1. These include the following:

- **Unknown:** A risk that has not yet been identified
- **Draft:** A proposed risk that has not yet been validated
- **Rejected:** A Draft risk that is not valid
- **Escalated:** A Draft risk that is outside the scope of the project and that should be managed elsewhere in the organisation
- **Active:** A valid risk with a probability/frequency greater than zero and that will impact one or more project objectives if it occurs. An Active threat can affect the project negatively, while an Active opportunity has a potential positive effect.
- **Deleted:** A risk that is no longer valid, perhaps resulting from a change in the project's strategy, environment, objectives, or scope.
- **Expired:** The time period in which the risk could have occurred ("impact window") has passed, so the risk no longer needs to be considered.
- **Closed:** A risk (threat) for which the response has been fully effective and the risk can no longer affect the project.
- **Occurred:** The risk has happened and the impact is being experienced.

Using these status values, we can describe the lifecycle of a typical individual project risk:

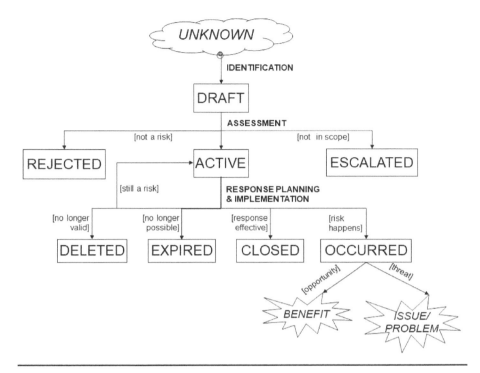

Figure 11-1  Risk Status Values

- All risks start as **Unknown**. When they are identified, they become **Draft** risks which need to be validated, because not everything identified during risk identification is in fact a risk.
- We need to filter the output of risk identification, and some Draft risks may be **Rejected** (for example, issues, problems, constraints, requirements, etc.). Rejected risks don't need to be recorded in the risk register, because they never were a risk to the project.
- We might also discover that some Draft risks are outside the scope of the project, because they wouldn't affect any of our objectives if they happened. But some of these could affect some other objective (for example, in another project, or a programme/portfolio, or ongoing business-as-usual activities or operations, or at a corporate or functional level). In these cases, we should **Escalate** the risk. We first determine who it should belong to, which is usually the person who owns the objective that would be affected if the risk occurred, and then we communicate details of the risk to that person. Once they have accepted responsibility for the

Escalated risk, we can remove it from our risk register, because it is not relevant to our project.
- Draft risks that are considered valid and in scope for the project become **Active**, and are recorded in the risk register.
- Active risks need to be assessed, and appropriate responses should be developed and implemented.
- After implementation of agreed risk responses, a number of outcomes are possible, so the status of Active risks should be monitored regularly.
  - Active risks may **remain Active** for some time. This means that an Active threat still has the potential to affect the project adversely, whereas an Active opportunity still represents a source of potential benefits.
  - Active risks might be marked as **Deleted** if they can no longer affect the project because of changes in the project context—for example, a scope change on the project, or adoption of different methods or technology. A Deleted risk remains in the risk register because it's important to have a record that it was recognised as a risk until the project context changed. Deleting a threat is a good outcome, whereas we don't want to see opportunities Deleted!
  - Active risks might become **Expired** if their impact window has passed. These remain in the risk register because they were relevant to our project at one time. We might be pleased that a threat has Expired, since it can no longer have a negative effect, but we might regret an Expired opportunity for which the positive impact is no longer possible.
  - When an Active threat is successfully managed so that it can no longer affect the project, it is marked as **Closed**. Opportunities cannot be Closed, because they will remain Active until they have either Occurred, Expired, or been Deleted.
  - If a risk actually happens, it is marked as **Occurred**. It is good for an opportunity to occur, and bad for a threat to occur. A risk might occur if the response to a threat proved ineffective or the response to an opportunity was successful (or perhaps it was due to chance or luck!). When a threat has Occurred, it is converted to an issue or problem and managed accordingly. When an opportunity has Occurred, the additional benefits must be recognised and managed.

We can record risk status values in the risk register and then we can use status metrics to monitor the effectiveness of the risk process. For example, as the project progresses, we can measure how many Draft risks become Active (indicating how well we identify real risks), and how many Active risks actually Occur or are Closed (showing how good our risk responses are). This should help us improve performance during the remainder of our project, and inform the risk process for future projects.

## Review risk response effectiveness

This involves the following elements:

- Consider whether planned risk responses are achieving the intended effect on individual threats and opportunities, and on the level of overall project risk. To do this, we need to assess the current probability/frequency and impacts for each individual risk and compare these with the predicted change that we expected when we developed the risk response. We also need to recalculate overall project risk by updating and re-running the quantitative risk analysis model.
- Assess whether any secondary risks have arisen as a result of implementing chosen risk responses.
- Identify where new responses are required to address existing risks, involving either additional risk actions to implement the chosen risk response strategy, or perhaps selecting a new risk response strategy.
- Confirm ownership of risk responses, recognising that the person originally nominated as risk owner may no longer be best placed to deal with the risk.

## Identify new risks

During the risk review, we need to consider whether any new risks have arisen since the last risk review, including secondary risks which result from risk actions that we've already implemented. We won't usually run another big risk identification exercise as part of the risk review step, unless this is a major project. Instead we might conduct a short brainstorm session, or conduct Delphi interviews with subject-matter experts, or use a nominal group technique or affinity analysis, or perhaps run through a risk checklist.

Any new threats or opportunities that we identify during the risk review need to be assessed and prioritised, then we must develop appropriate responses, and evaluate the effect on overall project risk. Data on new risks also need to be recorded in an update to the project risk register. This can all be done outside the risk review meeting, but it is important that we follow up on new risks that we find in the risk review, to ensure that they are included in the risk process and managed properly.

### Review risk process effectiveness

Having considered the status of existing risks, the effectiveness of current risk responses, and the existence of new risks, we should now have a good idea of whether the risk process is working as expected. If we're tracking risk metrics (as discussed in the previous chapter), these will also give us hard evidence of risk process effectiveness. We should be seeing the number (N) and average severity (A) of individual threats reducing, and the number and value of opportunities increasing. The Relative Risk Exposure Index (RRE) should be falling for threats and rising for opportunities. The level of overall project risk should also be moving in the right direction, with a higher probability of achieving project objectives and reduced variation in project outcomes.

By reviewing each of these indicators in the risk review meeting, we can determine whether the risk process as a whole is operating effectively. Where process performance is below expectations, we need to consider how to improve the risk process. This means revisiting the decisions we made when we launched the risk process, about the scope and parameters of risk management for our project (see Chapter 4). We may decide that the initial level of risk process was too simple for the size of risk challenge we actually face, or perhaps we overestimated the degree of risk in the project and we can now use a less rigorous risk process. We should consider whether we need to refresh or improve the process by using different techniques, changing the type of analysis undertaken, modifying reporting frequency, etc.

Any changes that we decide to make in the risk process as the result of the risk review will mean that the project's RMP needs to be updated, reissued, and communicated to all affected project stakeholders.

## SUMMARY AND REFLECTION QUESTIONS

It is the nature of risk to be dynamic, because the sources of uncertainty are constantly changing. The risk process needs to take this into account so that we can be alert and responsive to the evolving risk challenge. Having performed a single pass through the risk process, we can't rest on our laurels. In the world of risk, standing still is going backwards. Risk management is not a one-off exercise, but it must be repeated in order to maintain a current perspective on the risk exposure. To be fully effective, our risk process must be iterative, cyclic, proactive, and forward-looking. We should constantly be asking (and answering) questions such as:

- Where are we now?
- Which way are we heading?

- Are we still on course?
- What could affect achievement of objectives?
- What can be done about it?
- How have our actions changed things?
- Now where are we?
- etc. . . .

This is why we need to do risk reviews as part of the risk process, and why they are essential in providing an up-to-date view of risk exposure on our project.

## Reflection questions

- How often should you be conducting risk reviews on your current project? Why did you select this frequency? Are you doing this on your project (and if not, why not)?
- How can you know if the risk process is working effectively on your project? Is it?
- What should you do if the risk process appears to be ineffective?
- Are you monitoring changes in both threats and opportunities in your project?

## THE NEXT STEP ("NOW WE'VE REACHED A KEY POINT IN THE PROJECT, WHAT HAVE WE LEARNED?")

Chapters 4–11 have given us a complete approach to identifying and managing opportunities alongside threats within an integrated risk process. We now know how to set the scope of the risk process to include both types of risk, how to find opportunities, pick the winners, use numbers to model opportunities, decide what to do in addressing opportunities, take action, tell others, and keep it all up to date. If we follow this process throughout our project, we can minimise our threats and maximise our opportunities at the same time, giving our project the best chance of success. When our project comes to an end, we'll be able to look back and know that we've done everything possible to address all the uncertainties that matter. What a great feeling!

And it would be even greater if every project manager in our organisation felt the same. How can we share our experiences of including opportunity in the risk process with other colleagues? Maybe we shouldn't end the risk process without identifying and recording the lessons that we've learned along the way, and making those lessons available for others to learn before they embark on their own projects. There's one last step in the risk process which deals with precisely that challenge.

# Chapter 12
# Identifying Risk-Related Lessons

We might be forgiven for thinking that the risk process has come to an end. The previous chapter describes how to use risk reviews to keep our understanding of the risk exposure of our project up to date. We can just repeat those risk reviews throughout the lifetime of our project, and when the project ends, the risk process ends. Well, not quite. There's one more element of managing risk which we need to do, both throughout the project and at its end. We need to identify risk-related lessons.

## PURPOSE AND PRINCIPLES OF IDENTIFYING RISK-RELATED LESSONS

There are two reasons that organisations do projects. The first is to create project deliverables that will be used to deliver benefits and create value for our project stakeholders. The second reason is so that our organisation will know how to do similar projects better in future. Project-based organisations can use completed projects to create a body of knowledge and experience to benefit future projects. Unfortunately, most of us seem to forget this second aspect, so we lose out on at least half of the potential benefits of doing projects.

Project management processes usually include a step at the end for a post-project review, which is supposed to support organisational learning by capturing lessons from this project that can be applied to new similar projects.*  But post-project reviews are notoriously poorly performed, and sometimes not done at all. There are at least three reasons for this:

- Project teams usually disband as soon as the project ends, and people move on to new projects, taking their knowledge and experience with them.
- Some project sponsors see post-project reviews as an optional extra, and they may be reluctant to pay for something that doesn't directly benefit their own project.
- Knowledge management is an immature discipline in many organisations, so there's no infrastructure or process to capture and use previous experience.

When it comes to risk management, it's particularly important to learn from experience. Nobody likes making misteaks, and we certainly don't want to make the same misteak twice [sic], either in terms of being hit again by avoidable problems that we could have foreseen as threats, or by repeatedly missing potential benefits that we could have obtained if we'd identified them as opportunities. The only way to improve our practice of risk management is to identify and learn risk-related lessons. **The purpose of identifying risk-related lessons is to capture knowledge and experience in a form that can improve performance in remaining phases of this project and in future similar projects.**

Managing risk effectively is a difficult challenge, and we need all the help we can get. We don't have to start every project with a blank sheet of paper when it comes to risk management. If risk-related lessons have been identified by those who have gone before, we can build on their experience. And we in turn can record our own risk experiences, good and bad, for the benefit of those who come after us. That's the only way to improve the maturity of our risk management capability, both as individual project management professionals, and for project-based organisations. It also ensures that our organisations gain the full benefit from our project: not just through the deliverables that we produce, but in the risk-related lessons that we provide so that similar projects will be more likely to succeed in future.

Each previous chapter about the risk process has suggested that the stage being described is the most important. There are strong grounds for claiming that identifying risk-related lessons should take that place, because the value of these lessons extends beyond just our own project. If we identify and implement lessons from our recent experience into later phases of our own project, we,

---

* (Williams, 2007)

along with our project stakeholders, reap the benefits. But if lessons that we've identified lead to improved performance on future similar projects, the benefits are multiplied and ongoing. That's why this step in the risk process shouldn't be ignored, and it may even be the most important.

There are several key ***principles*** relating to identifying risk-related lessons.

## Identify lessons from both bad and good experience

Most of the time, when we think about what we need to do differently next time, our thoughts turn naturally to what went wrong and how to prevent it happening again, and how we (or someone else) could do better. This means that the majority of lessons we identify concern problems or issues that we encountered, and when we relate those to risks, we inevitably end up focusing on threats. Were there threats that we identified that we were unable to manage? Did we encounter problems that we could have foreseen as threats, and should we have been able to avoid or reduce them? Are there any generic threats that other projects should look out for, and did we find any particularly effective responses to them that might work for others?

While threat-related lessons are important, we mustn't forget opportunities. We should also reflect on what went well during our project, and where things were better than expected. In some cases, this will have been the result of identifying and capturing opportunities, and we should see if there are any generic opportunities that other projects might explore and effective opportunity responses that others could use.

## Do it regularly, don't wait until project completion

It's important to identify lessons about risk as soon as possible, so that we can use them in our own project as well as recording them to help future projects. We should, of course, hold a post-project review at the end of our project, but why wait? In addition to the post-project review, we should find ways to identify lessons about risk as our project proceeds, either in an ad hoc way, or in our risk review meetings, or by holding specific meetings at key points in the project (gateways, phase points, etc.).

If we only identify lessons at the end of the project, we deny ourselves the chance to benefit during the remainder of our own project. Why should the only ones to learn from our experience be other people? Let's learn from our own experience wherever possible!

## A lesson is not learned until it's implemented

We usually refer to things that we identify in this stage as "lessons learned". But just because we identify something doesn't mean that future practice will take it into account. Lessons are only truly learned when they have been implemented, either during the remainder of our project, or on a future similar project. As a result, we should really call them "lessons-to-be-learned". This emphasises the fact that something needs to change in our future behaviour and practice as a result of identifying a lesson.

## Record lessons in actionable form

If our focus is to identify lessons-to-be-learned, then we need to record each lesson in a form that can be used the next time a similar situation is encountered, either by us or by someone else. We must be specific in recording trigger conditions when the lesson might be relevant, as well as detailing actions that can be taken to avoid or minimise a threat, or to exploit or maximise an opportunity. We might be able to identify a particular occasion when this lesson would be relevant. Maybe we can nominate a person to take responsibility for ensuring that the learning from this lesson is embodied in process documentation of guidance notes, or the person who is most likely to be in a situation in which they need to put the lesson into practice.

If we enter our lessons into a corporate knowledge database, we should consider labelling each lesson with tags that would help others to query the database and find lessons that are relevant to them. Our organisation should develop a series of standard search terms that we can use, and there should also be a way of proactively notifying other project managers if lessons have been entered into the database that they should consider for their own projects.

## Summary

The purpose and principles of identifying risk-related lessons are summarised in Table 12-1.

**Table 12-1 Purpose and Principles of Identifying Risk-Related Lessons**

| Purpose | To capture knowledge and experience in a form that can improve performance in remaining phases of this project and in future similar projects |
|---|---|
| Principles | Identify lessons from both bad and good experience<br>Do it regularly, don't wait until project completion<br>A lesson is not learned until it's implemented<br>Record lessons in actionable form |

## TYPICAL TECHNIQUES FOR IDENTIFYING RISK-RELATED LESSONS

When we're looking for risk-related lessons, either during our project or at the end, there are a number of key inputs that we should review. These include:

- The most recent risk register, with information on all identified risks, including those that were active at the last review, as well as those that were previously marked as expired, occurred, closed, or deleted.
- The current risk report, which provides a commentary on changes in overall project risk exposure.
- The latest issue log or problem register (if we have one) describing problems that have occurred on our project, and hopefully indicating the root cause of each one. Some of these issues will result from threats that weren't properly managed, or opportunities that were missed, or secondary risks that arose from our risk responses.
- Project earned value data (if our project has used earned value analysis), which will show us where project performance has deviated from plan, either in schedule or cost (or both).
- The project change log (if present), which might show where risks have occurred and resulted in a significant change to the project (either positive or negative).

It's a good idea to structure our lessons-to-be-learned exercise in some way, and it's possible to identify lessons linked to each of our project phases. While this might make sense for our project, it may make it harder to transfer lessons to other projects if their phase structure is different. A good alternative is to use the Risk Breakdown Structure (RBS) as a framework. Many organisations will have a common RBS that they use for all similar projects, so linking risk-related lessons to the RBS allows projects similar to ours to find relevant lessons quickly.

We also need to make sure that the right people participate in our lessons-to-be-learned identification exercise. These should include as wide a range of project stakeholders as possible, but as a minimum the review should be attended by the project manager, key members of the project team, the project sponsor, and people who can represent the views of other key stakeholders.

Once we've decided how to structure our review and who should attend, we can identify risk-related lessons by asking the following questions:

- Threat-related questions:
  - What were the main threats identified on this project? Do any of these represent generic threats that might affect similar projects?

- Which foreseeable threats actually occurred, and why? What could we have done differently to prevent them occurring, or to make their impact less severe?
- Which issues or problems occurred that should have been foreseen as threats?
- Which responses were effective in managing threats? Can any of these be embedded into standard procedures?
- Which threat-focused risk responses were ineffective and why? Can they be improved, or should they not be used in future?
• Opportunity-related questions:
  - What were the main opportunities identified on this project? Do any of these represent generic opportunities that might affect similar projects?
  - Which opportunities that could have been captured were missed, and why? Are there actions we could have taken to exploit them, or to make them more beneficial?
  - Which unplanned benefits arose that should have been identified as opportunities?
  - Which responses were effective in managing opportunities? Can any of these be embedded into standard procedures?
  - Which opportunity-focused risk responses were ineffective and why? Can they be improved, or should they not be used in future?
• Risk process-related questions:
  - How much effort was spent on risk management, both to execute the risk process and to implement responses? Was this more or less than we had budgeted for risk management?
  - Can any specific benefits be attributed to the risk process—for example, reduced project duration or cost, increased business benefits or client satisfaction, etc.?
  - Can we calculate return on investment (ROI) for the risk process on our project, based on the cost of undertaking the risk process for this project, compared with the additional benefit obtained through avoided threats and exploited opportunities?
  - Which risk process elements worked well, and why? And which worked less well, and why? How could we improve our standard approach to risk management in future?
  - Which risk tools and techniques were most useful, and which were less valuable? How can we improve our use of tools and techniques in future?

As we answer these questions, we're aiming not only to record our experience, but to identify lessons from our project that could be useful to future similar projects, including:

- Generic risks (both threats and opportunities) that could affect future projects
- Risk responses and actions that have proved effective and should be actively considered for future projects
- Risk responses and actions that were tried but were ineffective, and which therefore might be excluded from future projects
- Secondary risks likely to arise from particular risk responses
- Elements of the risk process that were particularly effective or ineffective, and ways in which we could build on strengths and overcome weaknesses, including tips and hints on using risk tools and techniques

## WRITE IT DOWN

We need to record the findings of our lessons-to-be-learned exercise so that others can benefit from our insights. The main place to record our findings is in a lessons-to-be-learned register (the L2BL Register). An example format for this is shown in Figure 12-1.

In addition to the L2BL Register, we might also include a lessons-to-be-learned report as part of our risk communications. This might be issued as part of a larger project progress report (or post-project review report when we review lessons at the end of the project), or it might form a separate risk lessons report, depending on the reporting requirements of the project.

The report should include recommendations for the following:

- Individual threats and opportunities to be added to the organisation's risk checklist for consideration during the risk identification step of future similar projects
- Modifications to the organisation's RBS, if risks were identified that did not map into the existing RBS framework
- Proactive and preventative actions to be included in the strategy of future similar projects to address the types of threats and opportunities likely to be encountered
- Modifications to our standard project management approach that would avoid particular generic threats or exploit generic opportunities
- Changes to the risk process to improve effectiveness, either in use of tools or techniques, or in development of standard templates to support the process

Finally, if our organisation has a structured knowledge management system, we should ensure that the lessons we've identified are entered into it, with appropriate tags and data labels to enable others to search the knowledge base and find relevant lessons for their own situation.

LESSONS-TO-BE-LEARNED REGISTER FOR THE HARDSTONE PROJECT

PROJECT REF: KOBMP-6
SCOPE OF REVIEW: [project phase]
PERIOD UNDER REVIEW: From [date] to [date]
DATE OF REVIEW: [date]

| ID | TRIGGERING ACTIVITY OR EVENT | LESSON TO BE LEARNED | DATE IDENTIFIED | PERSON IDENTIFYING | LESSON OWNER responsible for implementation | NEXT IMPLEMENTATION OPPORTUNITY | EXPECTED IMPLEMENTATION DATE |
|---|---|---|---|---|---|---|---|
| KO-02 | Initial kick-off meeting had packed agenda, several team members felt that more time would have been valuable. | Consider extending duration of initial kick-off meeting from half-day to a full day for similar major projects. | 12-Mar-18 | DH | Programme Manager | Kick-off meeting for next major project | 01-Sep-18 |
| GD-03 | Some project team leaders are responsible for leading two development areas, as well as acting as reviewers for others. This created a very heavy workload. | Aim to make each project team leader responsible for just one development area, and reviewer for no more than two other areas. | 06-Apr-18 | GS | Resource Manager | Initiation phase of next similar project | 01-Sep-18 |
| TM-14 | The project manager has demonstrated exceptional skill in handling both difficult people and frustrating situations during the project, with grace and humour. | Exercise particular care when selecting the project manager for future high-profile strategic projects, to ensure that the selected person has the necessary people skills and emotional maturity to handle difficult people and situations. | 26-Jul-18 | MJS | Programme Manager | Initiation phase of next similar project | 15-Oct-18 |
| TM-22 | One key team member has left the company at short notice due to severe ill health, leaving several gaps in the roles to be covered by the team. | Ensure cross-training for key project roles in case someone leaves the team unexpectedly. Implement job-shadowing where possible. Identify a back-up for each key project role. | 09-Sep-18 | CSD | HR Department | Next review of staff training provision | 01-Mar-19 |
| PM-06 | The project has produced a good set of project governance documents. | Use governance docs from this project as templates for future projects. | 30-Nov-18 | DH | Project Control Office | Immediate | 05-Jan-19 |

**Figure 12-1** Example L2BL Register Format, with Sample Entries

## SUMMARY AND REFLECTION QUESTIONS

As we proceed through the various steps of the risk process in our project, we learn a lot about the uncertainties that could affect us, including individual threats and opportunities, and also factors that can affect the level of overall project risk exposure. This additional information should help us to manage risk more effectively during the remaining stages of our project, if we can turn it into practical lessons that we can act on.

But in addition to project-specific risk information, we also gain wider insights into how to make risk management work better in our organisation and with our types of project. These insights might come too late to help us in our own project, but they might assist others, so that they build on our successes and don't repeat our mistakes.

Identifying risk-related lessons-to-be-learned will help both us and those who come after us, if we can turn our experience into specific advice or guidance. The benefits of this learning are so significant that we can't afford to ignore this final step in the risk process.

### Reflection questions

- Why should we refer to outputs from this step as "lessons-to-be-learned" and not "lessons learned"?
- Did you have access to lessons from previously completed similar projects at the start of your current project, or during project execution? If yes, did this help? If no, what difference might it have made to your project?
- How would you persuade your project sponsor or client that identifying lessons-to-be-learned is not a waste of time or effort on this project?

## FINAL REFLECTION QUESTIONS

This chapter has outlined the importance of identifying lessons-to-be-learned as we go along, while they are fresh in our minds, and before we get distracted by the next challenge or activity on our to-do list. Perhaps we should follow our own advice, and do exactly that right now.

Hopefully as you've read the preceding chapters describing how to identify and manage opportunities throughout the risk process, you've had a number of "Aha!" moments, insights into new ways of thinking about risk and different ways to manage it effectively. Maybe you've been comparing the approaches we've described here with your own practice, and identifying areas where you could and should do things differently.

Table 12-2 Risk Process Steps—Purpose and Principles

| Process Step | Purpose | Principles |
|---|---|---|
| Risk management planning | To ensure that the risk approach on this project is appropriate and effective | Define objectives at risk and scope of risk process.<br>Reflect stakeholder risk appetite in measurable risk thresholds.<br>Tailor risk process to match the risk challenge. |
| Risk identification | To identify knowable risks that otherwise would not be managed | All risks are uncertain.<br>Each risk must be linked to at least one objective.<br>Consider different time perspectives.<br>Use more than one risk identification technique.<br>Include multiple perspectives.<br>Consider all potential sources of risk.<br>Repeat risk identification throughout the project. |
| Qualitative risk assessment | To evaluate key characteristics of identified risks in order to prioritise them for further attention and action | Use consistent and objective assessment framework.<br>Reflect corporate risk appetite and risk thresholds.<br>Seek input from different perspectives.<br>Beware bias. |
| Quantitative risk analysis (QRA) | To evaluate overall project risk by considering the combined effect of uncertainty on project outcomes | QRA is not always needed.<br>Where QRA is required, know why and how to use it.<br>No models are "correct".<br>Include all types of uncertainty.<br>Use best available data.<br>Use the results. |
| Risk response planning | To identify appropriate ways to address individual threats and opportunities, as well as ways to manage overall project risk | Strategy before tactics.<br>Deal equally with both threats and opportunities.<br>Respond to overall risk, not just individual project risks.<br>The first idea is not always best.<br>Responses must match the level of risk.<br>Be creative but realistic.<br>Ensure clear ownership. |
| Risk response implementation | To ensure that agreed risk response strategies and actions are implemented effectively | Don't cut corners.<br>Motivate action owners.<br>Assume nothing. |
| Risk communication and reporting | To provide project stakeholders with timely and accurate risk information to support appropriate risk-informed decision-making and action | Be honest.<br>Be specific.<br>Be timely. |
| Risk reviews | To provide visibility of current risk exposure | Decide when and where to review risk.<br>Review risk regularly.<br>Don't limit reviews to meetings.<br>Involve key stakeholders. |
| Identifying risk-related lessons | To capture knowledge and experience in a form that can improve performance in remaining phases of this project and in future similar projects | Identify lessons from both bad and good experience.<br>Do it regularly, don't wait until project completion.<br>A lesson is not learned until it's implemented.<br>Record lessons in actionable form. |

It would be a shame to lose those insights, so before moving on to the next part of the book, let's take time to review what we've learned so far about including opportunities in the risk process, identify our own lessons-to-be-learned, and work out how we're going to capture that learning and turn it into action and benefits. The following final reflection questions might help to shape your risk management thinking and practice in future:

- Table 12-2 summarises the purpose and principles of each step in the risk process. In your opinion, which step is the most important, and why?
- What were your "Aha!" moments from what you've read?
- Why is it important to include opportunities in the risk process?
- Which steps in the risk process are the most challenging or need the most modification when it comes to including opportunities?
- Does risk management in your organisation address both threats and opportunities in a single integrated risk process?
    - If yes, what benefits does this deliver to your project, to the organisation, and to your clients?
    - If no, what barriers might you face in your project or your organisation in introducing an integrated approach? How might you overcome these barriers?
- How would you describe a risk mindset that includes opportunities? Is this how you think about risk, or are there areas in which you need to modify your thinking?
- What three things will you do in the next three months to change how you think and act towards risk in your project?

# Section C
# And Finally...

# Chapter 13
# Making It Work

The aim of this book is help you find and capture upside risks in your projects. The opening chapters explained why opportunities are an integral part of the risk concept, and why they are important in the world of projects. After that we explored the various steps in a generic risk management process, looking at what might need to change if we want to include the management of opportunities.

Although process is doubtless important, we make a serious mistake if we think it is the whole story when it comes to managing opportunities in projects. Many organisations are convinced that all they need are the Three Ts: *Techniques, Tools,* and *Training,* and everything will be OK.

Of course, it's true that we do need a structured process for managing risk in projects, including both threats and opportunities, and this process involves use of a number of *techniques*. Many of these techniques require *tools* to support them, and people will need to be *trained* in how to use the techniques and tools effectively. There are many examples of organisations who developed a set of risk management procedures with a full range of techniques to identify, prioritise, quantify, and respond to project risks; who bought one or more software tools to support the risk process, and sent their project staff on risk training courses. Having invested significantly in these elements, they then sit back and expect risks to be managed effectively, with fewer problems and issues on their projects, more successful project delivery, motivated and happy project teams, satisfied stakeholders, and an increasingly profitable business.

It often comes as a rude surprise to these organisations when they discover that *the Three Ts are necessary but not sufficient.* Despite their well-structured set of techniques, supporting tools, and trained staff, many project-based organisations feel that they aren't getting the expected and promised benefits from risk management. Although they are implementing some form of project risk management, they still find their projects failing to achieve objectives—instead projects are late, over budget, or under-performing. It seems that although use of risk management is widespread, it doesn't always work in practice. Why is this? Is risk management just a lot of hype, or are there reasons for this shortfall?

The answer lies in a set of Critical Success Factors (CSFs) that lie outside the Three Ts.* These include:

- Common risk language
- Simple scaleable risk process
- Appropriate supporting risk infrastructure
- Strong and mature risk culture
- Organisational learning

These CSFs are important for effective risk management in general, but they are particularly relevant if we want to adopt an inclusive approach that covers both threats and opportunities in an integrated risk process. Each CSF is explored in turn in the following sections, focusing on how they apply to the broader application of risk management.

## CRITICAL SUCCESS FACTORS FOR MANAGING OPPORTUNITIES IN PROJECTS

CSFs are "critical" in two respects, both positive and negative. A CSF is critical because it must be present in order to make success possible. But it is also critical in the sense that if it is absent, success is impossible. Perhaps we should adopt a double interpretation of the CSF abbreviation, meaning both Critical Success Factor and Critical Source of Failure. In reality, success/failure is rarely binary, so the positive/negative influence of a CSF is more nuanced, as illustrated in Figure 13-1.

Taking this approach to CSFs, which factors must be present if we are to succeed in including opportunities alongside threats in an inclusive risk process, and which would inevitably lead to failure if they were absent? The CSFs listed above are discussed in the following sections, focusing on how they contribute (or not) to the broader application of risk management, remembering that each one is also important for effective risk management in general.

---

* (Hillson, 2002c, d, e, f)

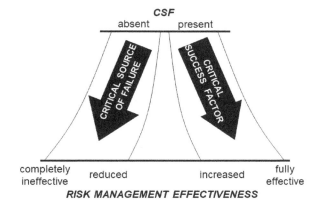

**Figure 13-1** Double Meaning of CSF

## Common risk language

The opening chapters of this book demonstrated that there is still an ongoing debate about the meaning of the term "risk". This matters because the definition we use to describe the concept of risk will inevitably shape the way we approach risk in practice. We use *language* to express our underlying *beliefs*, which in turn influence our *behaviour*.

- If we interpret "risk" as a synonym for "threat", then we will focus the risk process on finding, prioritising, and addressing only negative risks, and we will naturally exclude any consideration of the upside (although we may use a separate process to identify and capture opportunities, outside of risk management). When we attend risk workshops, our language will shape the discussion around *"What could go wrong?"* and *"How bad could it be?"* Response development will focus on *"How can we stop it, or at least make it less bad?"* (prevention or mitigation), and we will measure success by the absence of problems.
- Alternatively, if we adopt an inclusive view of risk that covers both threats and opportunities, or even a neutral risk definition that allows for the possibility of positive and negative outcomes from uncertainty, then our risk process is more likely to consciously seek out opportunities alongside threats, and develop proactive responses to both types of risk. We will use more general language in risk workshops and response development sessions, with questions such as, *"What uncertainties might affect us, for better or worse?"*, *"How significant might the impact be?"*, and *"How can we influence the uncertainty or its outcome?"* Success is seen in terms of both fewer problems and increased achievement of objectives.

A shared definition of risk is a CSF for effective risk management, because without it, participants in the risk process will not be working together towards a common goal. Indeed, if their views about risk differ fundamentally, they may even be pulling in opposite directions. Some may believe that "All risks are bad" and resist any attempt to look for potential upside risks. Others will be raising uncertainties that could have a positive impact on project success, and expecting these to be addressed within the risk process, perhaps even resenting the resistance of their threat-focused colleagues.

An agreed definition of risk is not a theoretical detail or an optional luxury. It is an essential prerequisite for a focused risk process wherein all interested parties share a common purpose. Before risk can be managed effectively, stakeholders and participants in the risk process need to agree on what "risk" means. But this doesn't only matter in general terms; it is specifically relevant to the inclusion of opportunities in the risk process:

- If project stakeholders lack a common risk language that recognises the dual nature of risk, it will not be possible for them to act coherently, and opportunities will not be identified and managed effectively (absence of a common risk language is a Critical Source of Failure).
- Conversely, if all are agreed that opportunities are an integral part of risk, we will pull together and ensure that they are not overlooked, but instead upside risk will receive the level of attention and action that it deserves (presence of a common risk language is a Critical Success Factor).

## Simple scaleable risk process

When it comes to the risk process, many organisations seem to believe that bigger is better. More risk techniques, more risk tools, more risk reporting—surely this must result in more management of risk? This often leads project managers and their teams to view risk management as a bureaucratic overhead, imposed by management to control the project, adding another layer of unnecessary analysis and reporting, but with little additional benefit. In order to facilitate effective risk management, the risk process must be *simple* enough to meet the needs of every project, and *scaleable* to address different project types, while covering the essential steps necessary to enable risk to be managed proactively and effectively.

The risk process described in earlier chapters is not "rocket science". Instead, it is merely structured common sense, as illustrated in Chapter 3 by the use of a logical set of questions to shape the process. One important CSF for effective risk management is to ensure that the risk process is as simple as possible but also scaleable to the needs of any project. Otherwise, risk management will be rejected by busy project managers as too difficult or onerous.

If this seems obvious, then it is even more so if we want to extend the risk process to include opportunities. Without a simple underlying risk process, identifying and prioritising opportunities as well as threats will be seen as increasing the effort consumed by risk management. If our teams have to develop and implement responses to capture or maximise upside risk, while already undertaking additional tasks to avoid or minimise threats, they might object to the imposition of "too much extra work". As a result, a simple scaleable risk process is a CSF for the successful implementation of twin-focused risk management addressing both threats and opportunities:

- If the risk process used to address threats is already viewed as a bureaucratic overhead that takes up too much time and effort for busy project teams, then attempting to add in the consideration of opportunities will surely be met by stiff resistance (absence of a simple scaleable risk process is a Critical Source of Failure).
- However, if the underlying risk process is seen as light-touch while still being effective, it will be easier for project teams to make the necessary process modifications to include opportunities, especially if they are convinced of the benefits that they can expect by proactively identifying and capturing upside risk (presence of a simple scaleable risk process is a Critical Success Factor).

## Appropriate supporting risk infrastructure

Most processes need some level of supporting infrastructure, including the Three Ts (techniques, tools, training), and the risk process is no exception. We've already seen that effective risk management requires a simple scaleable risk process, and scaleability means that we should expect the level of process detail to vary from one project to another. Low-risk projects may only need a simple risk process, whereas more challenging projects might require a more in-depth approach. The level of supporting infrastructure required is driven by the level of risk process being implemented. Each organisation needs first to choose a level of risk management implementation which is appropriate, acceptable, and affordable, and then provide the necessary infrastructure to support it.

At one extreme, we might adopt an informal risk process which includes all the phases, but with a very light touch. This might simply involve regularly asking and answering the set of structuring questions described in Chapter 3. At its simplest, the project manager might address the questions at the start of each week as part of a personal planning routine. Alternatively, the project team might be asked to consider the questions during regular project progress meetings. If the structuring questions are followed by action and repeated regularly, then the full risk process can be followed with no supporting infrastructure at all.

At the other extreme is a fully-detailed risk process that uses a range of techniques and tools to support the various phases. For example, we might hold a series of stakeholder workshops during project initiation in order to develop the Risk Management Plan (RMP). We could use multiple risk identification techniques involving a full range of project stakeholders singly and in groups, followed by both qualitative risk assessment (with a risk register and various structural analyses) and quantitative risk analysis (using Monte Carlo simulation, decision trees, or other statistical methods). Detailed risk response planning at both strategic and tactical levels might include calculation of risk-effectiveness, as well as consideration of secondary risks arising from response implementation.

Both the super-simple and the deeply-detailed risk processes are extremes, and most organisations will implement a level of risk management somewhere between these two. The two extremes do, however, illustrate how we can retain a common risk process that includes all the required steps, while selecting very different levels of implementation.

Having selected the level of *implementation*, we can then decide how to provide the required level of *infrastructure* to support the risk process. This might include choosing *techniques*, buying or developing software *tools*, providing *training* in both knowledge and skills, producing *templates* for various elements of the risk process, and considering the need for *technical support* from risk specialists. The required level for each of these infrastructure elements will be different depending on the chosen risk process implementation level.

Failure to provide an appropriate level of infrastructure can cripple risk management in an organisation. Too little support makes it difficult to implement the risk process efficiently, while too much infrastructure adds to the cost overhead. Getting the supporting infrastructure right is therefore a CSF for effective risk management, enabling the chosen level of risk process to deliver the expected benefits to the organisation and its projects.

Appropriate supporting risk infrastructure isn't only important for risk management in general. It is particularly necessary when we are including opportunities in the risk process. Taking each of the infrastructure elements mentioned above:

- *Techniques.* Although the underlying risk process is the same whether we are addressing only threats or considering opportunities as well, some modifications are needed to the standard risk techniques when we're dealing with opportunities, and there are also specific additional techniques that we might choose to use, as discussed in the earlier part of this book.

- *Tools.* Many current popular risk software tools don't have the functionality needed to include opportunities. We need risk software tools that cover both threats and opportunities, especially for more complex projects that require an in-depth risk process with more detailed analysis. This is discussed further below.
- *Training.* Many risk training courses still focus exclusively on threats, leaving project team members lacking in both the knowledge and skills required to address opportunities effectively.
- *Templates.* We can improve efficiency and consistency by developing templates for various elements of the risk process (risk management plan, risk register, interview outlines, checklists/questionnaires, Risk Breakdown Structure [RBS], risk reports, etc.). Many of these will need to mention opportunities explicitly if they are to support a broader risk process.
- *Technical support.* Sometimes we need to call in risk experts to assist us, especially for more complex projects that use specialist risk techniques or tools. These experts might come from another project or department, a Project Support Office, an internal risk centre of excellence, or an external consultancy. If our risk process includes opportunities, we'll need to ensure that the expertise of our chosen experts also covers this aspect of managing risk.

Appropriate supporting risk infrastructure is a CSF for a broader inclusive approach to risk management:

- If the risk infrastructure is threat-focused, our ability to identify, prioritise, respond to, and report on opportunities alongside threats will be severely compromised, and may be impossible (absence of appropriate supporting risk infrastructure is a Critical Source of Failure).
- Having risk techniques, tools, training, templates, and technical support available that support inclusive risk management will allow us to address opportunities efficiently (presence of appropriate supporting risk infrastructure is a Critical Success Factor).

## Strong and mature risk culture

Culture is a hot topic in academic, business, and political circles, and much has been written and said about it. A lot of the material on culture is hard to understand and not very practical, which makes it hard to apply in real situations. Leading academics and authors on culture disagree on its definition and characteristics, and various culture models have been developed in an attempt to crystallise the essential elements of culture.

Perhaps the simplest and most useful culture model is the A-B-C Model,[*] which has three elements:

- *Attitude* is the chosen position adopted by an individual or group in relation to a given situation, influenced by perception.
- *Behaviour* comprises external observable actions, including decisions, processes, communications, etc.
- *Culture* is the values, beliefs, knowledge, and understanding, shared by a group of people with a common purpose.

The A-B-C Model is based on the following considerations:

- Culture arises from repeated Behaviour.
- Behaviour is shaped by underlying Attitudes.
- Both Behaviour and Attitudes are influenced by the prevailing Culture.

This is relevant to the context of managing risk, because risk culture is a key influence on risk management effectiveness. To understand risk culture, we can produce a risk variant of the A-B-C Model[†] (Figure 13-2) simply by replacing its generic elements with risk-related versions, as follows:

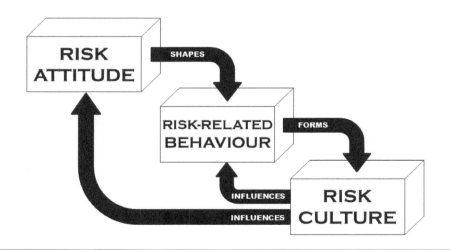

**Figure 13-2**   A-B-C Model for Risk

---

[*] (Hillson, 2013; Institute of Risk Management, 2012a, b)
[†] (Hillson, 2013)

- *Risk attitude* is the chosen position adopted by an individual or group towards risk, influenced by risk perception.*
- *Risk-related behaviour* comprises external observable risk-related actions, including risk-based decision-making, risk processes, risk communications, etc.
- *Risk culture* is the values, beliefs, knowledge, and understanding about risk, shared by a group of people with a common purpose.

The A-B-C Model for risk indicates that *risk culture* is formed by repeated *risk-related behaviour*, which in turn is shaped by our *risk attitudes*. There's also a dynamic feedback loop, with the prevailing risk culture influencing underlying risk attitudes and ongoing risk-related behaviour. A strong and mature risk culture is therefore a CSF for risk management effectiveness, because it affects the way we think about risk and behave towards risk.

Risk culture exists at all levels within an organisation, from top to bottom, and ideally it will be aligned and coherent across the business. There are two steps to developing a strong and mature risk culture:

- **Top-down.** Setting risk culture directly from the top requires a clear statement of intent from leaders in the organisation, laying out their vision and policy for risk management, describing their values and beliefs about risk, and explaining the approach that they intend to take in order to exploit risk and create benefits. The desired risk culture should be actively communicated to all staff, so no one is any doubt about how risk will be addressed within the organisation, and appropriate risk-related behaviour is actively promoted and encouraged.
- **Bottom-up.** Having defined the desired organisational risk culture, we can then allow risk culture to emerge naturally from within the organisation at each level. As people put risk management into practice within their routine jobs, they will start to experience fewer problems and enhanced benefits, and recognise the importance of managing risk. Their belief in the value of risk management will reinforce the correct behaviour. A positive cycle is created in which acting properly towards risk creates a strong risk culture, and that in turn encourages the right risk-related behaviour.

An organisation needs to be proactive if it wants to develop a strong and mature risk culture. Otherwise a risk culture will emerge that is based on whatever people happen to believe about risk and however they happen to behave in risky situations. There is no guarantee that such an unmanaged emergent risk culture will lead to effective management of risk.

---

* (Hillson & Murray-Webster, 2007)

Ideally, senior management will recognise the importance of risk culture, and take steps to define and develop the desired risk culture across the business. However, if our organisation is not addressing risk culture proactively from the top, all is not lost. We can take steps at project level to ensure that the local risk culture within our team is strong and mature, encouraging appropriate risk-related behaviour. The A-B-C Model reveals that effective action to change risk culture needs to start with attitudes towards risk. The A⇨B⇨C links indicate that if we can manage and modify risk attitudes, then those changed risk attitudes will influence risk-related behaviour in the right direction, and hence build a new and more mature risk culture. This means that adopting the right attitude towards risk is the first step in building a strong risk culture.[*]

This is where the A-B-C Model becomes relevant to the inclusion of opportunities in our approach to managing risk. If our risk attitude is threat-focused, it will encourage a set of threat-based risk-related behaviours which in turn build a risk culture that reinforces the threat-only approach. On the other hand, if our risk attitude includes the existence of upside risk, this will shape our risk-related behaviours to address opportunity alongside threat, and lead to a risk culture that supports the broader inclusive approach to risk management.

The characteristics of a threat-focused risk attitude are very different from one which recognises the importance of opportunities, as shown in Table 13-1.

It is easy to see how these different risk attitudes would encourage very different sets of risk-related behaviours, and hence build a different risk culture. The resulting risk culture in turn reinforces risk attitudes and risk-related behaviour, leading to a vicious or virtuous cycle. If we encourage inclusive attitudes towards risk among our project teams and other stakeholders, we can develop a strong and mature risk culture that supports inclusion of opportunities in the risk process. Such a risk culture is a CSF:

Table 13-1 Characteristics of Different Risk Attitudes

| Threat-Focused | Opportunity-Inclusive |
|---|---|
| Risk is avoidable. | Risk is natural. |
| Risk is always bad. | Risk can also be good. |
| "High risk" means dangerous. | "High risk" can be exciting & challenging. |
| All risk should be prevented. | Some risk should be exploited. |
| Risk management protects value. | Risk management enhances value. |
| Discussing risk shows weakness. | Discussing risk shows maturity. |

---

[*] (Hillson & Murray-Webster, 2007)

- An emergent risk culture based on whatever people happen to believe about risk and however they happen to behave in risky situations is unlikely to lead to effective management of opportunities through the risk process, as many people still adopt threat-focused attitudes and actions (absence of a strong and mature risk culture is a Critical Source of Failure).
- A managed risk culture based on opportunity-inclusive risk attitudes will support risk-related behaviour that addresses opportunities alongside threats (presence of a strong and mature risk culture is a Critical Success Factor).

## Organisational learning

Chapter 12 discussed the importance of identifying risk-related lessons-to-be-learned (L2BL) as an integral part of the risk process, to be conducted not only at the end of the project but also at key points during the project lifecycle. This allows our current project to benefit from our experience to date, ensuring that future phases don't repeat the mistakes of earlier phases. But in order to maximise the benefit of capturing risk-related L2BL from individual projects, we need to have a learning organisation that has a structured way to capture and communicate knowledge beyond a single project.

Knowledge management (KM) is a relatively new topic in the context of projects, and it doesn't yet feature strongly in project management bodes of knowledge, training courses, or practice. At the organisational level however, KM has been considered for a little longer, and many organisations have developed structured systems to create, capture, and communicate knowledge for the benefit of the whole enterprise. Supporting KM infrastructure is becoming more widely available, with knowledge repositories or databases forming the hub that allows practical implementation of KM.

Risk information should obviously form part of any KM system, based on the set of questions provided in Chapter 12, including:

- Generic risks
- Risk responses that have worked or not worked in the past
- Commonly encountered secondary risks
- Risk process elements that were particularly effective or ineffective

The purpose of recording this knowledge is to ensure that future projects are able to learn from past experience, improving the effectiveness of risk management going forward. The learning organisation should be continually improving its risk management effectiveness with time, because L2BL are actually learned, implemented, and integrated into routine practice. Without such learning, organisations are most likely to repeat past mistakes, and progress towards risk management maturity will be slow, if at all.

The ability to improve future risk management effectiveness based on past performance is clearly also relevant to the emerging practice of managing opportunities within the risk process. This will be new to many organisations, and early experience is likely to be mixed. Project teams seeking to implement an inclusive approach that covers both threats and opportunities will need all the help they can get, especially with the unfamiliar aspects of identifying, prioritising, and responding to opportunities, and access to L2BL will make their life a lot easier.

As a result, organisational learning that includes opportunity-related L2BL is a CSF:

- Without access to structured knowledge about managing opportunities, based on previous experience in related projects, each project will have to start from scratch when seeking opportunities and trying to turn them into benefits, reducing risk management effectiveness and making the project more likely to repeat past mistakes (absence of organisational learning is a Critical Source of Failure).
- The organisation that provides well-structured risk information about generic opportunities, effective responses, and ways to strengthen the risk process will give its project teams the best chance of identifying and capturing opportunities (presence of organisational learning is a Critical Success Factor).

## CSF summary

Each of the CSFs discussed above is an organisational factor that lies outside the influence or authority of the project manager. Together they form the context within which projects have to manage risk. Not only do these factors affect the general effectiveness of risk management in projects, but they have a direct influence on the success or failure of any attempt to include opportunities explicitly within the risk management approach. This is indicated by the dual interpretation of the abbreviation CSF to mean both Critical Success Factor and Critical Source of Failure (see Figure 13-1). Table 13-2 summarises why each of these CSFs matters if we are serious about adopting an inclusive approach to risk management.

Table 13-2 CSF Summary

| Factor | Critical Source of Failure | Critical Success Factor |
|---|---|---|
| Common risk language | If project stakeholders lack a common risk language that recognises the dual nature of risk, it will not be possible for them to act coherently, and opportunities will not be identified and managed effectively. | If all are agreed that opportunities are an integral part of risk, project stakeholders will pull together and ensure that they are not overlooked, and upside risk will receive the level of attention and action that it deserves. |
| Simple scaleable risk process | If the risk process used to address threats is already viewed as a bureaucratic overhead that takes up too much time and effort for busy project teams, then attempting to add in the consideration of opportunities will be resisted. | If the risk process is seen as light-touch while still being effective, it will be easier for project teams to make the necessary process modifications to include opportunities. |
| Appropriate supporting risk infrastructure | If the risk infrastructure is threat-focused, our ability to identify, prioritise, respond to, and report on opportunities alongside threats will be severely compromised, and may be impossible. | Having risk techniques, tools, training, templates, and technical support available that support inclusive risk management will allow us to address opportunities efficiently. |
| Strong and mature risk culture | An emergent risk culture based on whatever people happen to believe about risk and however they happen to behave in risky situations is unlikely to lead to effective management of opportunities through the risk process, because many people still adopt threat-focused attitudes and actions. | A managed risk culture based on opportunity-inclusive risk attitudes will support risk-related behaviour that addresses opportunities alongside threats. |
| Organisational learning | Without access to structured knowledge about managing opportunities, based on previous experience in related projects, each project will have to start from scratch when seeking opportunities and trying to turn them into benefits. | The organisation that provides well-structured risk information about generic opportunities, effective opportunity responses, and ways to strengthen the risk process will give its project teams the best chance of identifying and capturing opportunities. |

# REMAINING WORK

Although the case has been made that a wider scope for risk management is a logical and positive development, not all the pieces are yet in place to make the implementation effective and trouble-free. There are still some issues that require attention if risk management is to be broadened successfully to include opportunity. What is still needed to make it work in practice?

## Standards and guidelines

Significant progress has been made in the definitions of risk contained in the main risk management standards, as discussed in Chapter 1. Most of these now contain either explicitly dual definitions that cover both positive and negative sides of risk, or at least give a neutral definition that can be interpreted in the broader sense of threat and opportunity.

Unfortunately, progress has lagged behind somewhat in the work of many academics, authors, and providers of risk training. Although leading professional organisations in the project management field, such as the Project Management Institute (PMI®) and the Association for Project Management (APM), now emphasise that risk includes both upside and downside, this needs to be accepted and reflected more widely by those who provide practical risk guidelines. It is too common to find disclaimers near the beginning of academic papers or textbooks on risk that say, *"While the idea that risk includes opportunity as well as threat is suggested by some standards, this is not frequently adopted in practice by most organisations, so we'll concentrate here on managing real risks that can have an adverse effect on project success."* This merely perpetuates a limited understanding and practice of risk management and doesn't help individuals, project teams, or organisations to take full advantage of risk management by maximising and capturing opportunities as well as minimising and avoiding threats.

The definition debate has been settled in favour of a broader understanding of risk, as reflected in the majority of international risk standards, and those who provide practical guidelines, textbooks, or training need to produce materials that support this understanding.

## Process modifications

The underlying risk process applies equally whether we view risk management as exclusively threat-focused or whether we wish to cover both threats and opportunities. An inclusive risk process requires some process modifications which are simple extensions of the familiar threat-focused approach,\* and Part B (Chapters 4–12) has addressed these in some detail. There are, however, a number of new techniques or novel applications of existing methods which are recommended for inclusion in the risk process to assist in identifying, assessing, and managing opportunities.

---

\* (Hillson, 2002g; Olsson, 2006)

Novel additions to the risk management process suggested in previous chapters include the following:

- Risk management planning
  - A clear statement in the RMP that the risk process includes opportunities within its scope
  - Definition of project-specific impact scales to include positive impacts which would arise from the capture of opportunities, as well as the negative impacts of threats which might occur
- Risk identification
  - Extension of familiar techniques to look for upside uncertainties, including constraints analysis (based on assumptions analysis), benefit tree analysis (analogous to fault tree analysis), or pre-mortems (linked to futures thinking)
  - Use of "double-sided techniques" such as SWOT Analysis or Force Field Analysis
- Qualitative risk assessment
  - Use of the "mirror" Probability-Impact Matrix (P-I Matrix) to prioritise the worst threats and best opportunities for further attention and action
- Quantitative risk analysis
  - Explicit consideration of upside variability when generating input data for a Monte Carlo risk model
  - Use of stochastic branches to model key opportunities
- Risk response planning
  - Use of risk response strategy options for opportunities (Escalate, Exploit, Share, Enhance, Accept)
- Risk response implementation
  - Development of specific actions to implement opportunity strategies, to ensure that the good intentions of the risk response planning phase are translated into reality
  - Considering the possibility of secondary opportunities arising from implementation of responses to identified risks
- Risk communication and reporting
  - Modifying outputs from the risk process to include and communicate opportunity information, including the risk register, risk reports, metrics, etc.

Each of these process modifications needs to be embodied in the organisational procedures used for project risk management, and also included in risk skills training for project team members so that new opportunity-focused techniques can be used effectively. Procedure development will also be required so that new opportunity techniques interface seamlessly with existing threat-focused techniques and other project management and business processes. Opportunity elements of the risk process should be seen as a fully integrated part of the whole, rather than an optional extra to be implemented in special cases. The goal is for participants in the risk process to use all risk techniques equally and naturally, without considering whether a particular technique is relevant to threats or opportunities.

## Supporting risk software tools

We've seen that having an appropriate level of supporting infrastructure is a CSF for risk management effectiveness, and that it is also required if we want to include opportunities in the risk process. Although we may accept this in principle, in practice it is often difficult for an organisation to provide the infrastructure their projects need, because many of the currently available risk software tools lack the required functionality. This is not the case for all risk tools, of course, and it's encouraging to see a gradual increase in the capability of risk tools to support the broader risk process that includes both threats and opportunities.

The areas in which current risk tools seem to be most lacking include the following:

- *Risk register.* As we progress through the various steps in the risk process, we need to record data about individual project risks into a risk register, in which data fields have to cope with both opportunities and threats (see Table 4-2). This particularly affects the ability to record positive impacts for opportunities (savings in time, money or resources; increases in performance, reputation, safety, regulatory compliance), as well as negative impacts for threats (see Table 4-4). Where risk register software autogenerates risk scores (see Figure 6-3), these also need to work for both upside and downside risks.
- *Risk reporting.* Many elements of risk reports are generated automatically as outputs from a risk database. These include:
  ○ *Prioritised risk lists*—for example, a "Top 10" or "Top 20". We need to able to produce such lists that either include both threats and opportunities, or that present top (worst) threats and top (best) opportunities separately.
  ○ *Risk distributions.* We can categorise risks by various characteristics, such as their source (perhaps mapped into the RBS), area of impact

(mapped to WBS), proximity, urgency, etc. We should be able to produce risk distributions that include both upside and downside risk, or for each separately.
- *Probability-Impact Matrix.* It is increasingly common to use the double "mirror" format for a P-I Matrix that shows threats on one side and opportunities on the other (Figure 4-5). The P-I Matrix is usually produced automatically from the risk database that lies behind a risk register, and software tools need to be able to generate the double "mirror" format.
- *Risk metrics.* Where our risk process produces a risk score for each individual project risk (for example, see Figure 6-3), we can then use risk tools to create various risk metrics to track and predict changes in risk exposure (see Figures 10-5 and 10-6). Metrics need to be available for both threats and opportunities, with matching outputs.

- *Historical risk information.* Risk information from previously completed projects can be held in a variety of electronic formats, including checklists, questionnaires, generic prompt lists, a risk library or risk repository, risk knowledge bases, etc. Some risk software tools embed historical risk information, other tools provide stand-alone automated versions. In each case, it is essential that the format allows information on opportunities to be included, queried, and analysed. Clearly, where previous experience of identifying and managing opportunities has been limited, there will be less historical information available to be included in these tools. Hopefully, as use of the inclusive approach becomes more widespread, risk tools will develop to include information about opportunities.
- *Novel opportunity identification techniques.* Chapter 5 discussed several techniques that can be used to identify opportunities in projects, including SWOT analysis, assumptions and constraints analysis, force-field analysis, benefits trees, and futures thinking. There are very few software tools available to support the use of these techniques in the project risk process, and this could be an area of focused future attention for risk tools providers.
- *Quantitative risk analysis.* Chapter 7 discussed ways in which routine quantitative risk analysis methods can be used to model the effect of opportunities, including use of stochastic branches to represent possible beneficial outcomes (see Figures 7-10, 7-11, and 7-12). This functionality is often not supported by the Monte Carlo software tools, making it impossible to produce a realistic risk model that represents all types of uncertainty, both upside and downside. One of the standard outputs from Monte Carlo simulation is the tornado chart, and the format of this needs to be able to reflect those opportunities to which the project outcome is particularly sensitive, as well as threats.

- *Overall project risk exposure.* The concept of overall project risk (see Chapter 7) has been discussed for some time, but its use in practice has been very limited. This may change as professional risk guidelines increasingly begin to emphasise its importance. However, there are very few currently available risk tools which support the calculation of overall project risk exposure explicitly, other than standard Monte Carlo modelling software. In addition, the ability to quantify overall project risk exposure is required in order to calculate appropriate amounts of contingency, as well as enabling decisions on risk efficiency to be made at programme or portfolio levels. It is not possible to calculate overall project risk exposure accurately without including opportunities, so when tools are developed to support this important aspect of the risk process, they must be able to take account of upside risks.

Tool vendors are naturally driven by the needs of the market, so perhaps the lack of comprehensive functionality to support management of opportunities within the risk process simply reflects what risk practitioners currently ask for. Hopefully, as awareness increases, tool vendors will recognise the demand and respond positively.

## THE FUTURE FOR OPPORTUNITY MANAGEMENT

As we come towards the end of this exploration of how to find and exploit opportunities in projects, it's time to consider where this is all leading. What is the future for opportunity management? A lot has changed in the past 20 years since we started thinking seriously about how to broaden project risk management to include upside as well as downside risk. Where do we go next?

It seems that many of the pieces are now in place to support widespread adoption of inclusive risk management that addresses both threats and opportunities in projects. We have clear definitions of risk in international standards that include upside risk, increasingly supported by professional bodies, recognised thought-leaders, expert practitioners, and leading-edge organisations. All that's needed is for more people to start doing it! It's time to stop talking about managing opportunities in our projects, and start managing them for real!

There are several reasons that an organisation might decide to adopt an inclusive approach to project risk management that addresses opportunities alongside threats:

- *Compliance.* International risk standards use a double-sided definition of risk that includes both threats and opportunities, and organisations wishing to align with these standards will have to follow suit.

- *Contracts.* Some client organisations impose on their suppliers and contractors a contractual requirement for formal risk management, and those wishing to do business with such clients must be able to demonstrate that they have risk processes in place. Where the client includes opportunity management within the contractual requirement, the supplier or contractor must do likewise.
- *Conformance.* It is increasingly common for major organisations to use the risk process to identify and capture opportunities proactively as well as to deal with threats. This is particularly true of businesses at the leading edge of their industry, who might be seen as role models or "best of breed". This may influence other organisations to adopt a similar approach, in order to keep up with the leading players in their industry.
- *Confidence.* Fear of failure is a strong motivation for many senior leaders, and anything that offers the possibility of improved performance or increased success is likely to be grasped.

Each of these motivations is understandable, but not well founded. The best reason to implement an integrated risk approach that includes proactive management of both upside and downside risk is because it ***provides concrete benefits*** to the organisation and its projects, including the following:

- *More realised opportunities.* Including opportunity management within the risk process ensures that we spend structured time looking for opportunities and addressing them proactively. This means that opportunities that might have been missed can be tackled, and some of them will be captured.
- *Improved chances of project success.* As opportunities are identified and captured, so projects will gain the associated benefits which would otherwise have been missed, leading to more successful projects which achieve their objectives.
- *No new process.* The use of a common process for managing both threats and opportunities ensures maximum efficiency, with the same activities being used for both types of risk. There is no need to develop, introduce, and maintain a separate opportunity management process, because one risk process will suffice.
- *Cost-effectiveness* (double "bangs per buck"). The use of a single process to achieve proactive management of both threats and opportunities will result in avoidance or minimisation of problems AND exploitation and maximisation of benefits.
- *Extension from the familiar techniques.* Most of the techniques currently used to manage threats can be adapted with only minor changes to deal with opportunities. Additional opportunity-specific techniques are not complicated and should be simple to adopt.

- *Minimal additional training.* Because inclusive risk management uses the familiar risk process, staff will only need training in new techniques for identifying and capturing opportunities.
- *Better contingency management.* Inclusion of potential upside impacts as well as downside means that contingency calculations are likely to be more realistic, taking account of both potential overspend and possible savings.
- *Increased team motivation.* Encouraging people to think creatively about ways to work better, simpler, faster, more effectively, etc. is a great motivator, and teams will enjoy looking for opportunities and making them happen.
- *Enhanced professionalism.* Clients who see supplier or contractor organisations and project teams working to maximise the benefits on their project will be impressed at the display of professionalism, leading to increased reputation and business growth.

In the final resort, inclusion of opportunity management as an integral part of the risk process will only become widespread and accepted if there are demonstrable benefits. In theory, there are many such benefits, but these must be seen in practice if busy project managers and their teams or resource-stretched businesses are to accept the need to commit additional effort to managing risk.

The best way to encourage more widespread adoption of management of opportunity within the risk process is for those organisations at the leading edge in implementing this approach to demonstrate the benefits. When others see the advantages gained from the integrated approach, increasing numbers of them will choose to follow the trailblazers.

If there is to be a future for opportunity management in projects as part of an inclusive risk process, it must become normal routine practice, and not seen as something unusual or special. We've spent the past 20 years talking about whether and how to do it. Now it's time to act!

# References and Further Reading

## GENERAL REFERENCES AND FURTHER READING

Apgar, D. (2006). *Risk intelligence: Learning to manage what we don't know.* Boston (MA), USA: Harvard Business School Press.
Arnoldi, J. (2009). *Risk.* Cambridge, UK: Polity Press.
Association for Project Management. (2004). *Project Risk Analysis & Management (PRAM) Guide,* second edition. High Wycombe, Bucks, UK: APM Publishing.
Association for Project Management. (2012). *APM Body of Knowledge,* sixth edition. High Wycombe, Bucks, UK: APM Publishing.
Association for Project Management. (2019). *APM Body of Knowledge,* 7th edition. High Wycombe, Bucks, UK: APM Publishing.
Beck, U. (2009). *World at risk.* Cambridge, UK: Polity Press.
Bernstein, P. L. (1996). *Against the Gods—The remarkable story of risk.* Chichester, UK: J Wiley.
Borge, D. (2001). *The book of risk.* Chichester, UK: J Wiley.
Burgess, M. (1999). *Living dangerously: The complex science of risk.* London, UK: Channel 4 Television.
Chapman, C. B. & Ward, S. C. (1997). *Project risk management: Processes, techniques and insights.* Chichester, UK: J Wiley.
Chapman, C. B. & Ward, S. C. (2000). "Estimation and evaluation of uncertainty—A minimalist first-pass approach." *Int J Project Management* 18, no. 6, pp. 369-383.
Chapman, C. B. & Ward, S. C. (2002). *Managing project risk and uncertainty.* Chichester, UK: J Wiley.
Chapman, C. B. & Ward, S. C. (2012). *How to manage project opportunity and risk.* Chichester, UK: J Wiley.
Chapman, R. J. (2011). *Simple tools and techniques for enterprise risk management,* second edition. Chichester, UK: J Wiley.
Cleary, S. & Malleret, T. (2007). *Global risk: Business success in turbulent times.* Basingstoke, UK: Palgrave Macmillan.

Cooper, D. F., Bosnich, P., Grey, S., Purdy, G., Raymond, G., Walker, P., & Wood, M. (2014). *Project risk management guidelines: Managing risk with ISO 31000 and IEC*. Chichester, UK: J Wiley.

Edwards, P. J. & Bowen, P. A. (2005). *Risk management in project organisations*. Oxford, UK: Elsevier.

Flyvbjerg, B., Bruzelius, N., & Rothergatter, W. (2003). *Megaprojects and risk: An anatomy of ambition*. Cambridge, UK: Cambridge University Press.

Gardner, D. (2008). *Risk: The science and politics of fear*. London, UK: Virgin Books.

Hillson, D. A. (1998). "Project risk management: Future developments." *Int J Project & Business Risk Mgt* 2, Issue 2, pp. 181–195.

Hillson, D. A. (2003). *Effective opportunity management for projects: Exploiting positive risk*. Boca Raton (FL), USA: Routledge/Taylor & Francis.

Hillson, D. A. (ed.) (2007a). *The risk management universe: A guided tour* (revised edition). London, UK: British Standards Institution.

Hillson, D. A. (2009). *Managing risk in projects*. Farnham, UK: Routledge/Gower.

Hillson, D. A. (2010). *Exploiting future uncertainty: Creating value from risk*. Farnham, UK: Routledge/Gower.

Hillson, D. A. (2014a). *The Risk Doctor's cures for common risk ailments*. Vienna (VA), USA: Berrett-Koehler.

Hillson, D. A. (ed.) (2016a). *The risk management handbook: A practical guide to managing the multiple dimensions of risk*. London, UK: Kogan Page.

Hillson, D. A. & Murray-Webster, R. (2007). *Understanding and managing risk attitude*, (second edition). Aldershot, UK: Routledge/Gower.

Hillson, D. A. & Simon, P. W. (2012). *Practical project risk management: The ATOM Methodology*, (second edition). Vienna (VA), USA: Berrett-Koehler.

Institution of Civil Engineers, Institute and Faculty of Actuaries. (2014). *Risk Analysis & Management for Projects (RAMP)* third edition. London UK: ICE Publishing.

International Organization for Standardization. (2018). *ISO 31000:2018: Risk Management Guidelines*. Geneva, Switzerland: International Organization for Standardization.

Murray-Webster, R. & Hillson, D. A. (2008). *Managing group risk attitude*. Aldershot, UK: Routledge/Gower.

Office of Government Commerce (OGC). (2010). *Management of Risk: Guidance for practitioners*, third edition. London, UK: The Stationery Office.

Pritchard, C. (2014). *Risk management: Concepts and guidance*, fifth edition. Boca Raton (FL), USA: Taylor & Francis.

Project Management Institute. (2009). *Practice Standard for Project Risk Management*. Newtown Square (PA), USA: Project Management Institute.

Project Management Institute. (2017). *A Guide to the Project Management Body of Knowledge (PMBOK® Guide)*, Sixth Edition. Newtown Square (PA), USA: Project Management Institute.

Project Management Institute. (2019). *Standard for Risk Management in Portfolios, Programs and Projects*. Newtown Square (PA), USA: Project Management Institute.

Taleb, N. N. (2007). *The Black Swan: The impact of the highly improbable*. London UK: Allen Lane/Penguin.

Taylor, L. (2014). *Practical Enterprise Risk Management: How to optimize business strategies through managed risk-taking*. London, UK: Kogan Page.

Wideman, R. M. (1992). *Project and program risk management: A guide to managing risks and opportunities*. Newtown Square (PA), USA: Project Management Institute.

# SPECIFIC REFERENCES FOR CHAPTERS

## Chapter 1

Association for Project Management. (2012). *APM Body of Knowledge*, sixth edition. High Wycombe, Bucks, UK: APM Publishing.

Dawkins, R. (1989). *The selfish gene*, second edition. Oxford, UK: Oxford University Press.

European Commission Centre of Excellence in Project Management (CoEPM²). (2016). *PM² Project Management Methodology Guide—Open Edition*. Brussels, Belgium: European Commission.

Goleman, D. (1995). *Emotional intelligence: Why it can matter more than IQ*. London, UK: Bloomsbury Publishing.

Hillson, D. A. (2003). *Effective opportunity management for projects: Exploiting positive risk*. Boca Raton (FL), USA: Routledge/Taylor & Francis.

Hillson, D. A. & Simon, P. W. (2012). *Practical project risk management: The ATOM Methodology*, (second edition). Vienna (VA), USA: Berrett-Koehler.

Hofstede, G. H. (1982). *Culture's consequences: International differences in work-related values*, abridged edition. Newbury Park (CA), USA: Sage Publications Inc.

Hulett, D. T., Hillson, D. A., & Kohl, R. (2002). "Defining risk: A debate." *Cutter IT Journal* 15, no. 2, pp. 4–10.

Institution of Civil Engineers, Institute and Faculty of Actuaries. (2014). *Risk Analysis & Management for Projects (RAMP)* third edition. London UK: ICE Publishing.

International Organization for Standardization. (2018). *ISO 31000:2018: Risk Management Guidelines*. Geneva, Switzerland: International Organization for Standardization.

Institute of Risk Management (IRM). (2012a). *Risk culture: Guidance for boards*. London, UK: Institute of Risk Management.

Institute of Risk Management (IRM). (2012b). *Risk culture: Resources for practitioners*. London, UK: Institute of Risk Management.

Maslow, A. H. (1943). "A theory of human motivation." *Psychological Review* 50, no. 4, pp. 370–396.

Maslow, A. H. (1987). *Motivation and personality*. New York (NY), USA: Harper Collins.

Office of Government Commerce (OGC). (2010). *Management of Risk: Guidance for Practitioners*, third edition. London, UK: The Stationery Office.

Project Management Institute. (2017). *A Guide to the Project Management Body of Knowledge (PMBOK® Guide)*, Sixth Edition. Newtown Square (PA), USA: Project Management Institute.

Project Management Institute. (2019). *Standard for Risk Management in Portfolios, Programs and Projects*. Newtown Square (PA), USA: Project Management Institute.

## Chapter 2

Hillson, D. A. (2014b). "How risky is your project? And what are you doing about it?" Proceedings of the PMI Global Congress North America 2014, Phoenix (AZ), USA, 26 October 2014.

## Chapter 3

Association for Project Management. (2012). *APM Body of Knowledge*, sixth edition. High Wycombe, Bucks, UK: APM Publishing.
International Organization for Standardization. (2018). *ISO 31000:2018: Risk Management Guidelines*. Geneva, Switzerland: International Organization for Standardization.
Office of Government Commerce (OGC). (2010). *Management of Risk: Guidance for practitioners*, third edition. London, UK: The Stationery Office.
Project Management Institute. (2017). *A Guide to the Project Management Body of Knowledge (PMBOK® Guide)*, Sixth Edition. Newtown Square (PA), USA: Project Management Institute.

## Chapter 4

Association for Project Management. (2008). *Prioritising project risks*. High Wycombe, Bucks, UK: APM Publishing.
Hillson, D. A. (2002a). "The Risk Breakdown Structure (RBS) as an aid to effective risk management." Proceedings of the 5th European Project Management Conference (PMI Europe 2002), Cannes, France, 19–20 June 2002.
Hillson, D. A. (2002b). "Using the Risk Breakdown Structure (RBS) to understand risks." Proceedings of the 33rd Annual Project Management Institute Seminars & Symposium (PMI 2002), San Antonio (TX), USA, 7–8 October 2002.
Hillson, D. A. (2012). "How much risk is too much risk? Understanding risk appetite." Proceedings of the PMI Global Congress EMEA 2012, Marseille, France, 8 May 2012.
Hillson, D. A. (2016b). "My stakeholders are my biggest risk." Proceedings of the PMI Global Congress NA 2016, San Diego (CA), USA, 27 September 2016.
Hillson, D. A. & Murray-Webster, R. (2011). "Using risk appetite and risk attitude to support appropriate risk-taking: A new taxonomy and model." *J Project, Program & Portfolio Management* 2, no. 1, pp. 29–46.
Hillson, D. A. & Murray-Webster, R. (2012). *A short guide to risk appetite*. Aldershot, UK: Routledge/Gower.

## Chapter 5

Hillson, D. A. (2000). "Project risks—identifying causes, risks and effects." *PM Network®* 14, no. 9, pp. 48–51.
Hillson, D. A. & Murray-Webster, R. (2007). *Understanding and managing risk attitude*, (second edition). Aldershot, UK: Routledge/Gower.
Klein, G. (1998). *Sources of power: How people make decisions*. Cambridge (MA), USA: MIT Press.
Klein, G. (2007). "Performing a project premortem." *Harvard Business Review*, September 2007, pp. 1–2.
Lewin, K. (1951). *Field theory in social science*. New York (NY), USA: Harper & Row.
Project Management Institute. (2009). *Practice Standard for Project Risk Management*. Newtown Square (PA), USA: Project Management Institute.
Williams T. M. (1994). "Using the risk register to integrate risk management in project definition." *Int J Project Management* 12, pp. 17–22.

## Chapter 6

Barber, R. B. (2003). "A systems toolbox for risk management." Proceedings of the ANZSYS Conference, Monash, Australia. November 2003.

Hillson, D. A. (2007b). "Understanding risk exposure using multiple hierarchies." Proceedings of the PMI Global Congress 2007 EMEA, Budapest, Hungary, 15 May 2007.

Hillson, D. A., Rafele, C., & Grimaldi, S. (2006). "Managing project risks using a cross Risk Breakdown Matrix." *Risk Management: An International Journal* 8, no. 1, pp. 61–76.

Kahneman, D., Slovic, P., & Tversky, A. (eds.) (1986). *Judgment under uncertainty: Heuristics and biases*. Cambridge, UK: Cambridge University Press.

Rafele, C., Hillson, D. A., & Grimaldi, S. (2005). "Understanding project risk exposure using the two-dimensional Risk Breakdown Matrix." Proceedings of the PMI Global Congress 2005 EMEA, Edinburgh, UK, 25 May 2005.

Slovic, P. (1987). "Perception of risk." *Science* 236, pp. 280–285.

Tversky, A. & Kahneman, D. (1974). "Judgment under uncertainty: Heuristics and biases." *Science* 185, pp. 1124–1131.

## Chapter 7

Box, G. E. P. & Draper, N. R. (1987). *Empirical model-building and response surfaces*. New York (NY), USA: John Wiley.

Box, G. E. P., Hunter, J. S., & Hunter, W. G. (2005). *Statistics for experimenters*, 2nd edition. New York (NY), USA: John Wiley.

Einstein, A. (1934). "On the method of theoretical physics." *Philosophy of Science* 1, no. 2, pp. 163–169.

Hillson, D. A. (2014b). "How risky is your project? And what are you doing about it?" Proceedings of the PMI Global Congress North America 2014, Phoenix (AZ), USA, 26 October 2014.

Hulett, D. T. (2009). *Practical schedule risk analysis*. Farnham, UK: Routledge/Gower.

Hulett, D. T. (2011). *Integrated cost-schedule risk analysis*. Farnham, UK: Routledge/Gower.

Vose, D. (2008). *Risk analysis—A quantitative guide*, third edition. Chichester, UK: J Wiley.

## Chapter 8

Hillson, D. A. (1999). "Developing effective risk responses." Proceedings of the 30th Annual Project Management Institute Seminars & Symposium, Philadelphia (PA), USA, 11–13 October 1999.

Hillson, D. A. (2001). "Effective strategies for exploiting opportunities." Proceedings of the 32nd Annual Project Management Institute Seminars & Symposium (PMI 2001), Nashville (TN), USA, 5–7 November 2001.

Hillson, D. A. (2002g). "Extending the risk process to manage opportunities." *Int J Project Management* 20, no. 3, pp. 235–240.

Maslow, A. H. (1943). "A theory of human motivation." *Psychological Review* 50, no. 4, pp. 370–396.

Maslow, A. H. (1987). *Motivation and personality*. New York (NY), USA: Harper Collins.

Project Management Institute. (2009). *Practice Standard for Project Risk Management*. Newtown Square (PA), USA: Project Management Institute.

Project Management Institute. (2017). *A Guide to the Project Management Body of Knowledge (PMBOK® Guide)*, Sixth Edition. Newtown Square (PA), USA: Project Management Institute.

## Chapter 9

Association for Project Management. (2004). *Project Risk Analysis & Management (PRAM) Guide* second edition. High Wycombe, Bucks, UK: APM Publishing.
Association for Project Management. (2012). *APM Body of Knowledge*, sixth edition. High Wycombe, Bucks, UK: APM Publishing.
Feldman, M. L. & Spratt, M. F. (1998). *Five frogs on a log: A CEO's field guide to accelerating the transition in mergers, acquisitions and gut wrenching change.* London, UK: Harper Business.
Office of Government Commerce (OGC). (2010). *Management of Risk: Guidance for practitioners*, third edition. London, UK: The Stationery Office.
Project Management Institute. (2017). *A Guide to the Project Management Body of Knowledge (PMBOK® Guide)*, Sixth Edition. Newtown Square (PA), USA: Project Management Institute.

## Chapter 10

Bourne, L. & Weaver, P. (2016). "Managing stakeholder risk." In *The Risk Management Handbook: A practical guide to managing the multiple dimensions of risk*, edited by Hillson, D. A. London, UK: Kogan Page.
Bourne, L. (2009). *Stakeholder relationship management: A maturity model for organisational implementation.* Aldershot, UK: Routledge/Gower.
Bourne, L. (2015). *Making projects work: Effective stakeholder and communication management.* Boca Raton (FL), USA: Routledge/Taylor & Francis.
Hillson, D. A. (2004). "Measuring changes in risk exposure." *The Measured* 4, Issue 3, pp. 11–14.
Hillson, D. A. (2011). "Enterprise Risk Management: Managing uncertainty and minimising surprise." In *Advising upwards*, edited by Bourne, L. Farnham, UK: Routledge/Gower.
Project Management Institute. (2017). *A Guide to the Project Management Body of Knowledge (PMBOK® Guide)*, Sixth Edition. Newtown Square (PA), USA: Project Management Institute.

## Chapter 11

[none]

## Chapter 12

Williams, T. (2007). *Post-project reviews to gain effective lessons learned.* Newtown Square (PA), USA: Project Management Institute.

## Chapter 13

Hillson, D. A. (2002c). "Critical Success Factors for effective risk management, Part 1: Agreed terminology." *Project Management Review*, July 2002, pp. 24–25.

Hillson, D. A. (2002d). "Critical Success Factors for effective risk management, Part 2: A simple process." *Project Management Review*, September 2002, p. 29.

Hillson, D. A. (2002e). "Critical Success Factors for effective risk management, Part 3: The right level of support." *Project Management Review*, October 2002, p. 17.

Hillson, D. A. (2002f). "Critical Success Factors for effective risk management, Part 4: Risk culture." *Project Management Review*, November 2002, p. 23.

Hillson, D. A. (2002g). "Extending the risk process to manage opportunities." *Int J Project Management* 20, no. 3, pp. 235–240.

Hillson, D. A. (2013). "The A-B-C of risk culture—How to be risk-mature." Proceedings of the PMI Global Congress North America 2013, New Orleans (LA), USA, 28 October 2013.

Hillson, D. A. & Murray-Webster, R. (2007). *Understanding and managing risk attitude,* (second edition). Aldershot, UK: Routledge/Gower.

Institute of Risk Management (IRM). (2012a). *Risk culture: Guidance for boards*. London, UK: Institute of Risk Management.

Institute of Risk Management (IRM). (2012b). *Risk culture: Resources for practitioners*. London, UK: Institute of Risk Management.

Olsson, R. (2006). *Managing project uncertainty by using an enhanced risk management process*. Västerås, Sweden: Mälardalen University.

# Index

## A

A-B-C Model, 242–244
accept, 164–166, 168–170, 249
action owner, 61, 63, 163, 175–177, 181–187, 230
action planning, 160, 169, 176
action window, 60, 108, 120–122, 124, 174, 177, 185, 203
Active Threat & Opportunity Management (ATOM), 26
affinity analysis, 217
ambiguity, 19, 20, 34, 133, 134, 136–138, 147, 152, 157, 158
analytical outputs, 136, 156
APM. See Association for Project Management
ARS. See Average Risk Score
Association for Project Management (APM), 7, 34, 35, 49, 248
assumptions analysis, 84–86, 97, 249
assumptions and constraints analysis, 87, 251
ATOM. See Active Threat & Opportunity Management
attention and action, 5, 20, 103, 104, 106, 111, 116, 122, 125, 126, 129, 130, 148, 177, 186, 200, 230, 238, 247, 249
Average Risk Score (ARS), 114
avoid, 19, 20, 25, 31, 69, 81, 164–167, 169, 170, 171

## B

BBS. See Benefits Breakdown Structure
Benefits Breakdown Structure (BBS), 117, 118, 203
benefit tree analysis, 88, 97, 249, 251
best opportunities, 46, 72, 101, 122–126, 158, 161, 177, 191, 200, 201, 209, 249, 250
beta distribution, 138
body of knowledge, 7, 26, 35, 49, 180, 221
brainstorm, 45, 84, 89, 91, 96, 217
bubble chart, 114, 115, 122
business-as-usual risks, 76, 77, 101

## C

cause-and-effect diagram, 84

cause, risk, and effect, 60, 81, 82, 87, 95, 96, 100
CBR. *See* cost-benefit ratio
CBS. *See* Cost Breakdown Structure
change log, 225
checklists, 45, 78, 83–85, 217, 227, 241, 251
cognitive bias, 76, 106
common causes, 92, 116, 118, 126, 129, 145, 168, 201
common process, 29, 45–48, 253
common risk language, 236–238, 247
communication plan, 198
compliance, 250, 252
conditional branch, 138, 139, 143, 153
confidence level, 150, 157
conformance, 253
connectedness, 107, 108, 116, 122
constraints analysis, 86, 97, 249
contingency, 8, 25, 62, 131, 168, 203, 252, 254
contract, 85, 86, 164, 167, 191, 253
controllability, 107, 108, 122
correlation, 133, 136, 137, 144–146, 148, 152, 153, 158
correlation coefficient, 145, 148
cost-benefit ratio (CBR), 172, 173
Cost Breakdown Structure (CBS), 60, 61, 117, 118, 122, 124, 129, 135, 203,
cost-effectiveness, 161, 171, 253
creativity techniques, 78
critical path, 108, 130, 133, 135, 148, 149
Critical Source of Failure, 236–247
criticality analysis, 146, 148, 156
criticality index, 149
Critical Success Factor, 236–247
culture, 17–21, 23–25, 97, 241–245
cumulative probability density function, 146. *See also* S-curve

current risk exposure, 211–213, 230

D

Dawkins, Richard, 24
deficiency needs, 22
definition debate, 8–10, 13, 16, 248
definitions of probability and impact, 61
Delphi technique, 84, 217
dependency, 133, 145, 152, 158
discrete distribution, 138
document review, 84, 85
double-sided concept of risk, 8, 18, 27, 39, 252

E

earned value, 18, 225
education, 23, 98
Einstein, Albert, 132
emotional intelligence, 23, 25, 26, 99
enhance, 165–170, 249
entrepreneurs, 27
escalate, 164–169, 214–216, 249
exploit, 46, 165–167, 169–171, 177, 211

F

facilitated workshop, 91, 96
facilitation, 62, 89, 96, 99
facilitator, 62, 64, 65, 78, 97, 105, 106
fault tree analysis, 84, 88, 97, 249
fear of failure, 19, 253
five frogs, 186
FMEA, 84
force-field analysis, 46, 91, 93–95, 97, 251
frequency, 63, 68, 69, 80, 107–109, 111–114, 122

future-focused techniques, 84, 89
futures thinking, 43, 84, 90, 97, 249, 251

**G**

generic risk, 227, 245
generic risk responses, 104, 117, 119
growth needs, 22

**H**

heuristics, 106
histogram, 147
historical data, 78, 251
Hofstede, Geert, 19
hotspots of risk exposure, 116, 126

**I**

impact, 6, 7, 33, 40, 45–47, 55, 63, 65–72, 83, 96, 107–116, 118–126, 145, 146, 153, 157, 159, 161, 164–170, 249, 250, 254
impact scale, 63, 68–71, 113, 249
impact window, 60, 108, 120, 121, 124, 173, 174, 203, 210, 214, 216
inclusive risk process, 13, 16, 23, 210, 236, 248, 254
individual risks, 34, 35, 40, 41, 44, 116, 130, 132–134, 138, 141, 144, 156, 159, 175–177, 182, 187, 195, 208, 209, 214, 217, 251
influence diagram, 84, 85, 135
initial risk assessment, 213
innovation, 25, 27
input data, 133, 134, 136, 137, 146, 149, 151, 153, 156, 249
integrated risk model, 135
integration, 47, 83
international risk standards, 6, 8, 13, 16, 17, 23, 27, 48, 49, 248, 252

interview, 80, 84, 89, 90, 96, 97, 187, 217, 241
Ishikawa diagram, 84
issue log, 225
iteration, 136, 138, 141, 146, 149

**K**

Key Risk Indicator (KRI), 114
KM. *See* knowledge management
knowledge database, 224, 251
knowledge management (KM), 44, 222, 227, 245
KRI. *See* Key Risk Indicator

**L**

L2BL. *See* lessons-to-be-learned
L2BL register. *See* lessons-to-be-learned register
L2BL report. *See* lessons-to-be-learned report
law of unintended consequences, 174
leadership tone, 20
learning organisation, 245
legacy processes, 22
lessons learned, 224, 229
lessons-learned database, 83
lessons-to-be-learned (L2BL), 224, 225, 227–229, 231, 245, 246
lessons-to-be-learned register (L2BL register), 227, 228
lessons-to-be-learned report (L2BL report), 227
likelihood, 63, 68, 69, 125, 133, 136, 138, 153
lognormal distribution, 138

**M**

manageability, 60, 107, 108, 114, 115, 122, 124

mapping risks, 118
Maslow, Abraham, 21, 22, 26, 161
maturity, 222, 244, 246
megaproject, 42–44, 54
memetics, 23–25
metrics, 194, 199, 203–206, 216, 218, 249, 251
mindset, 25, 26, 89, 90, 97–99, 101, 231
modified triangular distribution, 137
Monte Carlo, 135–151, 153, 157, 240, 249, 251, 252
motivation, 21, 26, 181, 182, 184, 253, 254
multiple perspectives, 79, 80, 105, 230

## N

national culture, 19, 20
nine structuring questions, 40–47
nominal group technique, 84, 217
non-risks, 32, 81, 82
non-risk uses of opportunity, 32
normal distribution, 138

## O

OBS. *See* Organisational Breakdown Structure
opportunity identification techniques, 84–96, 251
Organisational Breakdown Structure (OBS), 61, 117, 118, 203
organisational learning, 221, 236, 245–247
overall project risk, 34, 35, 41, 44, 46, 130, 131, 133, 134, 136, 149–151, 157–161, 163, 164, 168, 169, 175–177, 180, 182–185, 187, 189, 190, 194, 195, 207–209, 211, 217, 218, 225, 229, 230, 252

## P

past-focused techniques, 83, 84
PBS. *See* Product Breakdown Structure
persistence, 23, 24
P-I Matrix. *See* Probability-Impact Matrix
PMI®. *See* Project Management Institute
post-project review, 78, 83, 84, 221–223, 227
post-project review report, 78, 227
preferred risk response strategy, 169, 171, 173–176, 178
pre-mortem, 90, 97
present-focused techniques, 83, 85, 88
prioritisation zone, 70, 109, 111, 113, 118, 122, 124
prioritised risk list, 200, 250
prioritising risks, 41, 55, 61, 101, 104, 107–116, 120–123, 126, 173
probabilistic branch, 139–142, 144, 153–156
probability, 7, 43, 46, 54, 55, 59–61, 63, 65–71, 107–109, 111–114, 116, 118, 122, 124–126, 129, 138, 139, 141, 144, 146, 149, 150, 153, 157, 164–170, 177, 201, 211, 214, 217
Probability-Impact (P-I) Score, 200
Probability-Impact Matrix (P-I Matrix), 46, 61, 70–72, 107, 109–114, 122–124, 169–171, 201, 249, 251
probability scale, 67, 68
problem register, 225

process modifications, 239, 247–250
product breakdown structure (PBS), 117, 118, 201, 203
professional bodies, 10, 180, 252
professionalism, 190, 254
project characteristics, 29, 30, 117
project design, 29, 31
project environment, 29, 31, 32, 93
project hierarchies, 61, 117, 122, 124, 129
Project Management Institute (PMI®), 7, 34, 35, 49, 248
project manager, 20, 34, 167, 175, 178, 185, 190, 191, 193, 225, 246
project opportunities, 30, 33, 36, 39, 53, 88, 122, 167
project progress meeting, 212, 213, 240
project progress report, 227
project-specific impact scales, 249
project-specific risk information, 229
project sponsor, 34, 35, 54–56, 62, 64, 65, 79, 105, 131, 150, 151, 175, 178, 193, 194, 198, 222, 225, 229
project team culture, 20
prompt list, 79, 80, 84, 251
propinquity, 107, 108, 122
proximity, 60, 107, 108, 116, 120, 122, 124, 173, 174, 251
psychology, 17, 21–23

## Q

QRA. *See* quantitative risk analysis
qualitative risk assessment, 49, 59, 60, 103–127, 129–131, 141, 157, 161, 199, 201, 230, 240, 249
quantitative cost risk analysis, 135
quantitative risk analysis (QRA), 49, 59, 127, 129–158, 161, 176, 194, 195, 217, 230, 240, 249, 251
quantitative schedule risk analysis, 135
questionnaire, 84, 89, 96, 241, 251

## R

RACI (Responsible, Accountable, Consulted, Informed) analysis, 198
RBM. *See* Risk Breakdown Matrix
RBR. *See* risk-benefit ratio
RBS. *See* Risk Breakdown Structure
realised opportunities, 36, 253
real risks, 6, 77, 79, 81, 99, 101, 127, 180, 216, 248
reduce, 164–167, 170
Red/Yellow/Green, 60, 61, 70, 72, 73, 111, 114, 123, 124, 177, 201
relationship between threats and opportunities, 33
Relative Risk Exposure Index (RRE), 114, 204, 218
return on investment (ROI) for the risk process, 226
risicare, 8
risk action, 61–63, 169, 176, 177, 180–187, 203, 212, 213, 217
risk appetite, 55, 56, 59, 63–66, 74, 105, 106, 230
risk assessment criteria, 63, 65–70, 74, 105
risk assessment framework, 70–73, 105
risk attitude, 243–245, 247
risk-aversion, 19
risk-based decision-making, 134, 243
risk-benefit ratio (RBR), 173
Risk Breakdown Matrix (RBM), 118, 119

Risk Breakdown Structure (RBS), 57, 58, 61, 74, 79, 80, 117–119, 122, 124, 129, 201, 225, 227, 241, 250
risk categorisation, 61, 116–121
risk challenge, 42, 45, 53, 55, 56, 73, 99, 129, 179, 208–210, 218, 230
risk communication and reporting, 189–208, 230, 249
risk communications process, 198
risk culture, 19, 20, 236, 241–245, 247
risk database, 59, 250, 251
risk definition, 3, 6–16, 23, 32, 35, 237, 238, 248
risk distribution, 194, 195, 199, 201, 202, 250, 251
risk-effectiveness, 171–173, 184, 240
risk efficiency, 252
risk guidelines, 248, 252
risk identification, 49, 54, 60, 75–101, 103, 105, 162, 210, 211, 215, 217, 227, 230, 240, 249
risk identification techniques, 40, 43, 46, 78–80, 85, 89, 90, 93, 101, 230, 240
risk identification workshop, 80, 96
risk information, 41, 44, 78, 159, 190–195, 198, 199, 207, 229, 230, 245–247, 251
risk infrastructure, 236, 239–241, 247
risk library, 251
risk lifecycle, 214–217
risk list, 100, 104, 125, 195, 196, 200, 250
risk management effectiveness, 242, 243, 246, 250
Risk Management Plan (RMP), 56–63, 67, 70, 73–75, 99, 105, 107, 109, 117, 123, 124, 130, 131, 135, 198, 212, 218, 240, 241, 249

risk management planning, 49, 129, 230, 249
risk management process, 10, 26, 27, 35, 36, 39, 40, 44, 45, 53, 75, 78, 81, 101, 159, 162, 178–180, 186, 189, 191, 199, 209, 235, 249
risk management software, 26
risk management standards, 180, 248
risk metalanguage, 81–83, 86, 92, 95, 101
risk mindset, 231
risk model, 131–139, 141, 144–149, 151–153, 157, 158, 249, 251
risk owner, 61–63, 78, 99, 108, 118, 163, 175–177, 181–187, 200, 203, 213, 217
risk prioritisation chart, 114, 116, 122
risk priority, 61, 201
risk process effectiveness, 214, 216, 218
risk process steps, 230
risk profile, 174
risk register, 34, 59–63, 81, 99, 100, 103, 124, 133, 141, 153, 156, 161, 164, 166, 176, 177, 184, 185, 191, 192, 194, 195, 199, 200, 207, 215–217, 225, 240, 241, 249–251
risk report, 59, 62, 78, 156, 176, 184, 185, 190–192, 194, 195, 197, 199, 201, 207, 225, 241, 249, 250
risk reporting, 46, 59, 78, 191, 198, 201, 238, 250
risk repository, 251
risk response development, 111, 134
risk response effectiveness, 217
risk response implementation, 49, 179–187, 230, 249
risk response planning, 49, 78, 127, 157, 159–178, 182–184, 230, 240, 249

risk response strategy, 61, 134, 156, 160, 163–166, 168–178, 180–182, 187, 213, 217, 230, 249
risk review, 44, 49, 60, 100, 184, 209–219, 223, 230
risk score, 60, 108, 111–114, 116, 118, 122–125, 172, 173, 177, 250, 251
risk software tools, 241, 250, 251
risk status, 57, 60–62, 99, 100, 168, 177, 184, 185, 187, 194, 195, 198, 210, 211, 213–216, 218
risk thresholds, 40, 55, 56, 62–65, 68–70, 74, 105, 106, 158, 160, 164, 208, 209, 230
risk tool , 59, 132, 226, 227, 238, 250–252
risk-related behaviour, 20, 243–245, 247
risk-related lessons, 46, 49, 221–229, 230
risk-related roles and responsibilities, 55, 59, 62, 63
risks and risk, 34, 35
risk-seeking, 19
risky projects, 22, 29–32, 39
RMP. *See* Risk Management Plan
RMP template, 56, 57, 73
RRE. *See* Relative Risk Exposure Index

## S

scaleable risk process, 236, 238, 239, 247
scaling factors, 131
scenario analysis, 84, 90, 97
scope of (the) risk process, 54–57
S-curve, 146, 147, 149, 150, 156. *See also* cumulative probability density function

secondary risk, 33, 44, 61, 76, 171, 174, 175, 177, 184–186, 211, 217, 225, 227, 240, 245
sensitivity analysis, 144, 146, 148
serendipity, 47, 98
share, 165, 167–170, 249
simple project, 42, 44, 54, 57, 67, 157
sources of risk, 32, 79, 80, 117, 119, 195, 230
spike distribution, 138
stakeholder, 30, 31, 35, 40, 41, 43–46, 54–56, 59, 62–65, 69, 73, 76, 79, 104, 107, 108, 117, 118, 120, 167, 170, 175, 178, 186, 187, 189–195, 198, 199, 203, 207, 209, 213, 218, 221, 223, 225, 230, 235, 238, 240, 244, 247
stakeholder risk information needs analysis, 193–195
status of risk action, 185
stochastic branch, 133, 136, 138, 152, 153, 249, 251
strategic impact, 107, 108, 116, 119
strategy before tactics, 160, 163, 169, 230
strategy into actions, 176
subjectivity, 104
survey, 9–16, 27, 39
SWOT Analysis (Strengths, Weaknesses, Opportunities, Threats), 46, 91–93, 97, 249, 251
synergy, 31, 47, 137
system dynamics model, 85

## T

tailoring, 42, 45, 53–56, 59, 73, 75, 99, 117, 189, 191, 192, 194, 198, 209, 230
technical support, 240, 241, 247

template, 56, 57, 73, 227, 240, 241, 247
threat-focused, 13, 16, 22, 25, 26, 28, 48, 89, 90, 136, 168, 226, 238, 241, 244, 245, 247, 248, 250
three-point estimate, 60, 61, 124, 137, 138, 177. *See also* triangular distribution
Three Ts, 235, 236, 239
time-dependent [elements of the risk process], 59, 63, 73, 74, 210, 227, 240, 241, 250
timeliness, 171, 173, 198
time perspectives [for risk identification], 78–80, 82, 230
time window, 201, 203
timing, 119, 120, 122, 124, 173
top-priority risks, 70, 115, 132
"top risk" list, 114, 125, 194, 195, 200, 201
"Top 10" risks, 125, 200, 250
tornado chart, 148, 156, 251
Total Risk Score (TRS), 144
training, 23, 83, 87, 101, 135, 235, 239–241, 245, 247–249, 254
transfer, 164, 165, 167, 170
trend analysis, 194, 195, 199, 204
triangular distribution, 137, 141. *See also* three-point estimate

TRS. *See* Total Risk Score
two-dimensional techniques, 90, 101

**U**

uncertainty that matters, 3–7, 32, 33, 35, 36, 40, 54, 66, 77, 99, 107, 175, 219
uniform distribution, 138
urgency, 60, 107, 108, 114, 115, 120, 122, 124, 173, 174, 251

**V**

variability, 33, 34, 36, 133, 134, 136–139, 141, 147, 148, 152, 157, 158
visualisation, 84, 90, 97

**W**

WBS. *See* Work Breakdown Structure
wei ji, 8
Work Breakdown Structure (WBS), 18, 60, 61, 117–119, 122, 124, 129, 137, 203, 251
work package, 117, 119